Tending Your FOREST

A GUIDE TO ECOLOGICAL FOREST STEWARDSHIP

in the Eastern and Central United States

Paul Catanzaro and **Anthony D'Amato**

The mission of Storey Publishing is to serve our customers by publishing practical information that encourages personal independence in harmony with the environment.

EDITED BY Hannah Fries
ART DIRECTION AND BOOK DESIGN BY Erin Dawson
TEXT PRODUCTION BY Jennifer Jepson Smith

COVER PHOTOGRAPHY BY © Glass and Nature/Shutterstock .com, front t.l.; Jaakko Kemppainen/Unsplash, spine; Norm Eggert Photography, front t.r.; Paul Catanzaro, back; © Tibor Duris/Shutterstock.com, front b.

INTERIOR PHOTOGRAPHY BY © Adella Catanzaro, 35, 54, 138, 150, 154, 221; © Alexandra Kosiba, 119; © andreswd/iStock .com, 246; Anthony D'Amato, 5 inset, 14, 21, 27, 33, 34, 41, 56, 77, 78, 80, 112, 115, 118, 175, 184, 189, 196, 212, 214, 226; Connecticut Agricultural Experiment Station Archive, Connecticut Agricultural Experiment Station/© Bugwood.org /CC-BY-3.0-US/Wikimedia, 69 l.; Eli Sagor, 228, 241; © Evgeny Belenkov/iStock.com, 61 r.; © Heiko119/iStock .com, 69 t.r.; © Historic Collection/Alamy Stock Photo, 52; © Ivan kevin/Shutterstock.com, 263; Jennifer Fish, 37, 61 l., 65, 96, 217, 220, 230, 239; Courtesy of Jesse Bellemare, 53 r.; © Jon Benedictus/Shutterstock.com, 164 l.; Joshua Morse, 200; Courtesy of Julie OConnor, 177; © Kristin Pineda/ Shutterstock.com, 47; © L.A. Nature Graphics/Shutterstock .com, 58; Lina Denaroso, 39; Courtesy of Mass Audubon, 105; MassWildlife/Bill Byrne, 149, 160, 162; © maximmmmum /istock.com, 19 and throughout, background; Courtesy of Michael Grinley, 204, 211; © Mircea Costina/Shutterstock .com, 2; The U.S. National Archives (NWDNS-44-PA-2557) /Wikimedia Commons, 209; © NayaDadara/Shutterstock .com, 48; Norm Eggert Photography, 3, 11, 46, 87, 244; © ozgurdonmaz/iStock.com, 252; Paul Catanzaro, 94, 235; Courtesy of Peg Coleman, 224; © Petr Baumann/Shutterstock .com, 28; Courtesy of Raúl Chiesa, 89, 203; © Robert Winkler/ iStock.com, 18; Courtesy Sally Hightower, 109; Sean Bowers, 134; © Shades of Burgundy/Shutterstock.com, 5 main; Courtesy of Shawn Fraver, 22, 62; © Stefan Schottleitner/iStock .com, 164 r.; Susan M. Campbell, 6, 12, 92, 218; Tim Stout, 16, 25, 90, 158, 180; Courtesy of Tom Lautzenheiser/Mass Audubon, 72; © Tomasz Klejdysz/iStock.com, 69 b.r.; Courtesy of Town of Conway Historical Commission, 53 l.; © Troy Gipps, 171–172; © valleyboi63/Shutterstock.com, 101; Courtesy of Vermont Land Trust, 198; © Wirestock Creators/Shutterstock .com, 75

ILLUSTRATIONS AND DIAGRAMS BY © Paul Catanzaro
ADDITIONAL ILLUSTRATIONS BY © bsd studio/Shutterstock .com, 107; Courtesy of David Fox/TNC, 133; © Elena_Titova /Shutterstock.com, 15; Olivia Lukacic, 31; © Valedi /Shutterstock.com, 20 and throughout (border)

Storey Publishing
210 MASS MoCA Way
North Adams, MA 01247
storey.com

Storey Publishing is an imprint of Workman Publishing, a division of Hachette Book Group, Inc., 1290 Avenue of the Americas, New York, NY 10104. The Storey Publishing name and logo are registered trademarks of Hachette Book Group, Inc.

ISBNs: 978-1-63586-858-6 (paperback); 978-1-63586-859-3 (ebook)

Printed in China by Toppan Leefung Printing Ltd. on paper from responsible sources
10 9 8 7 6 5 4 3 2 1

TLF

Library of Congress Cataloging-in-Publication Data on file

PRAISE FOR *TENDING YOUR FOREST*

"Catanzaro and D'Amato have a gift for presenting complex biological information in an engaging and easy-to-read way that deepens our understanding of forests as complex, ever-changing ecosystems. With a focus on ecological forestry, *Tending Your Forest* shows us a path toward active stewardship that can both meet our material needs and help our forests adapt and thrive into the future."

—**ELI SAGOR, PhD,** Cloquet Forestry Center, University of Minnesota

"*Tending Your Forest* presents an innovative approach to stewarding the woods—one built on reciprocity between people and forests. As a forest ecologist, I'm thrilled to read a book that captures the nuances of ecological forestry in a format accessible to every landowner."

—**ZANDER EVANS,** executive director, Forest Stewards Guild

"Catanzaro and D'Amato illustrate how management can synchronize with ecological processes, bringing insight and solutions for healthy and resilient forests."

—**PETER J. SMALLIDGE,** New York State extension forester and director of the Arnot Teaching and Research Forest, Cornell University

"A much-needed resource for ecologically minded family forest owners, showing that forestry based on ecological principles provides a solid foundation for responsible management."

—**BRIAN PALIK,** forest ecologist, American Nature Solutions, and retired senior scientist, USDA Forest Service

"With chapters dedicated to climate, old growth, wildlife, resilience, economics, and more, this book serves as a guide to help get woodland owners more deeply connected to and engaged with caring well for their woods."

—**ALLYSON MUTH, EdD,** director of the James C. Finley Center for Private Forests at Penn State

"If landowners, land managers, or students of forestry and natural resources were to read only one book on ecological forestry, this should be it."

—**BARRIE BRUSILA,** consultant forester, Mid-Maine Forestry

"Paul Catanzaro and Anthony D'Amato have done an enormous service to foresters, landowners, and the land itself. By not simply defining ecological forestry but showing what it looks like in practice, Paul and Tony are connecting knowledge and wisdom to manage multiple challenges and truly practice ecological forestry. We cannot think of a greater need or a better pair to address it."

—**BRAD HUTNIK & GREG EDGE,** Wisconsin forest ecologists/silviculturists, hosts of the *SilviCast* podcast

"This indispensable, crystal-clear primer will empower the region's millions of private forest landowners to make vital choices for our forests' futures while also deepening their relationship to the land. In writing this guide, leading authorities Catanzaro and D'Amato have given a gift to generations of stewards."

—**JACKSON SAUL,** executive director of the Center for Northern Woodlands Education

"The definitive owner's manual for forest stewardship. D'Amato and Catanzaro offer a wealth of information and guidance relevant to the interests, challenges, and opportunities of the day. It deserves a space on the bookshelf of anyone who cares about forests."

—**STEVE HAGENBUCH,** senior forest program manager, National Audubon Society

"This thoughtful guidebook will give landowners a new framework for understanding forest dynamics and how to better care for their land."

—**LES BENEDICT,** assistant director, Saint Regis Mohawk Tribe

We dedicate this book to our friend, colleague, and mentor, David Kittredge.

Contents

Preface

I TOOK TONY'S OFFICE FROM HIM. He was finishing his doctorate at the University of Massachusetts, and I was starting my new job as an extension forester. Space is always very limited at universities. In order for me to have an office, someone had to lose theirs. The department head decided *that someone* was Tony. It was an awkward start to a relationship, to say the least.

Rather than a grudge, however, what began was a friendship and partnership. Although our friendship was born out of a common love of music, sports, and greasy diners, our partnership has been driven by a common belief that silviculture—*the science-based art of establishing, growing, and regenerating forests to meet landowner goals and the needs of society*—is more than growing timber like crops. We believed, and still do, that in silviculture lies the opportunity to restore forests and redefine our relationship with them.

It's been more than two decades since the office-stealing incident of 2004. Tony has gone on to a career as a silviculture professor, first at the University of Minnesota and then at the University of Vermont, where he now directs the forestry program. His work has focused on topics such as climate change and invasive insects and associated management strategies that address their impacts; perhaps most importantly, in partnership with Brian Palik of the USDA Forest Service, Tony has helped codify and advance the concepts of ecological forestry, bringing it from theory to practice.

My career took a different path. I immersed myself in helping to inform the decisions of family forest owners. My research and extension work has concentrated on topics such as conservation-based estate planning and forest stewardship. To bring more focus to the crucial role of private landowners, I collaborated with Dave Kittredge (University of Massachusetts) and Brett Butler (USDA Forest Service) to establish the Family Forest Research Center.

Though the paths that Tony and I took have been different, they have been highly complementary. We have identified critical issues facing forests and forest owners today and addressed them by providing digestible, science-based information. Our chosen delivery method became topic-focused hard-copy publications. Given the ownership pattern of our region, our target audience was and continues to be forest landowners, foresters, conservation organizations, and municipal leaders—the people who have the greatest influence on decision-making about our forested landscapes. Working with this audience has become our sweet spot and our professional joy.

Over the last couple of decades, we have produced a number of publications for this important audience. Our first publication, written while Tony was still a doctoral student, focused on the hot topic of the day in Massachusetts and beyond: high-grade harvesting, the removal of the highest-commercial-value trees with little or no effort directed at the well-being of the forest. Our second and third publications about restoring old-growth characteristics were based on Tony's dissertation work and, at Dave Kittredge's urging, strove to translate ideas that had their origins in the Pacific Northwest to the family-owned forests of our region. As critical issues arose, we took them on. Our outreach publications have focused on forest resilience, forest carbon, and the importance of responsibly addressing our tremendous amounts of daily wood consumption through reduced consumption and local production—all of which are covered in this book.

Each publication we wrote and each critical issue we wrestled with became a new piece of the puzzle. We began to see how these issues relate to one another and the way in which ecological forestry addresses them. The fuzzy outlines of ideas about the role of silviculture that we sensed early in our careers have come into sharper focus. However, since each of these publications was, by necessity, confined to a single topic, the connections among the publications were lost. This book provides us with the opportunity to connect the dots. It gives us the space to lay out foundational knowledge, critical issues, and strategies that weave the various topics together.

Information alone isn't enough, though. Through our outreach and applied work, we have seen that sustaining forests and people is more than just understanding the science of forests. A successful strategy must include both technical information and a healthy relationship with the forest. We believe ecological forestry offers an opportunity to cultivate and expand this new relationship. So, while this book does provide scientific information, it also includes an emphasis on the importance of our relationship with the forest and one another to help build a long-term strategy to sustain both forests and humans.

Finally, thank you. Simply by owning forested land and keeping it forested, you are providing tremendous public benefits. We're grateful for all you do. We hope this book helps grow your relationship with your forest and all those connected to it.

Paul Catanzaro

(Written in my timber-framed home in front of a woodstove burning local firewood marked by a consulting forester and harvested by a logger with whom I go to church)

Introduction

Ask a landowner what their goals are for their woods, and a common response is: "I just want to do the right thing and for my forest to be healthy." But what is the "right thing"? What does it mean for your forest to be "healthy"?

Shel Silverstein's book *The Giving Tree* provides an example of a one-way relationship with nature that is neither doing the "right thing" nor "healthy":

> And after a long time
> the boy came back again.
> "I am sorry, Boy,"
> said the tree, "but I have nothing
> left to give you—"

In his seminal book, *A Sand County Almanac*, Aldo Leopold calls for a different relationship with nature through the development of a land ethic, which does not separate people from the land but treats people and land as one community, each deserving equal respect and consideration. The quotes below from Leopold help articulate his land ethic.

> We abuse land because we regard
> it as a commodity belonging to us.
> When we see land as a community
> to which we belong, we may begin
> to use it with love and respect.

> That land yields a cultural
> harvest is a fact that is long
> known but latterly forgotten.

However, we would be sorely mistaken if we assumed that Leopold was the only or even the first person to describe a healthy relationship with the land. Indigenous peoples provide important examples of pathways to a different relationship with land. Robin Wall Kimmerer's *Braiding Sweetgrass* does a beautiful job describing it:

> A harvest is made honorable when it sustains the giver as well as the taker.

> Action on behalf of life transforms. Because the relationship between self and the world is reciprocal, it is not a question of first getting enlightened or saved and then acting. As we work to heal the earth, the earth heals us.

Ecological forestry is about strategies to steward forests and our relationship to them and one another.

Whether it's the words of Leopold, Kimmerer, or perhaps someone else altogether that resonate with you, consider the possibility that forest health is something more than the number of dying trees or the abundance of songbirds in your woods. Forest health includes the relationship between forests and people—a reciprocal, mutually beneficial, respectful relationship.

People have always relied on forests for survival and well-being. However, there is a strong case to be made that forests have never been more important than they are now. We have an increasing population of people relying on forests for an increasing amount of benefits. At the same time, we have a decreasing number of forests, and those forests are facing more and more challenges. Climate change, invasive species, biodiversity collapse, excessive herbivory, and forest conversion are all critical threats to our forests and, in turn, our society. Our response should not only address the threats before us but must also strive to put into place long-term solutions. As a forest landowner, you are the very foundation of these long-term solutions.

BUILDING A RELATIONSHIP

The challenge is to implement strategies across our forested landscapes that not only meet the personal and financial goals of individual landowners and their families but also maintain the public benefits that forests provide, all while putting into practice a relationship with our forests that will sustain both forests and people in the future.

Many people are drawn to forests for their valuable offerings, but relatively few have the chance or take the time to move beyond the superficial into a deeper relationship. Your forest may be where you have raised your family, where your children built forts and fairy houses. Your house may be warmed by wood from your own forest or perhaps your neighbors' forest. As a landowner you have an opportunity to see your forest over changing seasons and across years. You can watch it grow and change on its own. You are able to implement a stewardship strategy—an action for the long-term well-being of the forest—and observe how your actions change the

A landowner talks with foresters about his goals and forest stewardship options.

forest and how it changes you. There is no question that, as a landowner, you stand in a position to develop a deeper and more meaningful relationship with the forest.

Anyone who has been married for any length of time knows the real marriage doesn't happen on the day of the ceremony but rather on a daily basis through the routine of life. Who is cooking dinner and who is doing the dishes? How will you handle finances? Will you have children, and if so, how will you raise them? This is the day-to-day practice of marriage. Through this practice, a deep connection is developed and a commitment to one another is made over and over again. Such a relationship is hard work, and none of us is perfect. We're not going to hit the mark every day, but we need to do so on more days than not. When a marriage is healthy, the rewards are tremendous. Each person is not only made whole through the relationship, but the whole is greater than the sum of its parts. Such is our relationship with forests. It takes practice and a regular recommitment to the relationship.

Though ecological forestry is described in detail in Chapter 4 and throughout Part Two of this book and serves as the framework for the stewardship strategies in that section, we view ecological forestry as something more than nuts-and-bolts instructions for stewarding forests. We believe that embedded within ecological forestry is a relationship of reciprocity, a land ethic that honors the forest.

Embedded within ecological forestry is a relationship of reciprocity, a land ethic that honors the forest.

There are both passive and active strategies to steward forests and restore them. Our work to restore our forests is, in fact, work to restore our relationship to forests. It is an act of giving back, acknowledgment of the many benefits we receive. Ecological forestry expands the circle of our human ethics to include forests. It is an act of faith in our forests and their ability to renew and fully express themselves. It is also an act of hope in ourselves to move into the future with a different perspective and approach.

THE PAGES AHEAD

Part One of this book focuses on the essential ecological principles for understanding the ways in which forests establish, develop, and change. We also discuss the ways in which humans have altered forests, impairing their natural function. In the second part of the book, we will focus on reaching landowner goals through ecological forestry strategies to address these impairments, restore the missing parts of the forest, and increase forest resilience. Part Two will use the ecological principles described in Part One as the foundation of these strategies.

Chapters 6 through 12 each close with a "Practicing Ecological Forestry" box, which sums up specific aspects of ecological forestry that are important to understand within the context of that chapter's subject. You'll also find plenty of "fieldwork" prompts throughout that suggest opportunities to get out into your own woods and apply the concepts.

As testament to the crucial role of private landowners, we have included case studies of peers across the region that demonstrate the practice of ecological forestry. Find yourself within their stories. Consider their approaches and their applicability to your own goals and circumstances. Most important, let the stories inspire and empower you. Landowners are out in the landscape practicing ecological forestry, and you can, too.

It's our hope that after reading this book you will never look at a forest in the same way. This new way of seeing a forest will help you understand its past, your relationship to it, and opportunities to work with your forest to achieve your goals, serve the common good, and ensure the forest's health and resilience. You have a critical role to play, and the benefits are undeniable.

Ecological forest stewardship promotes all parts of the forest. Pictured are a trillium and squirrel corn growing at the base of a sugar maple, which is supporting moss communities on its bark.

THE REGION COVERED IN THIS BOOK

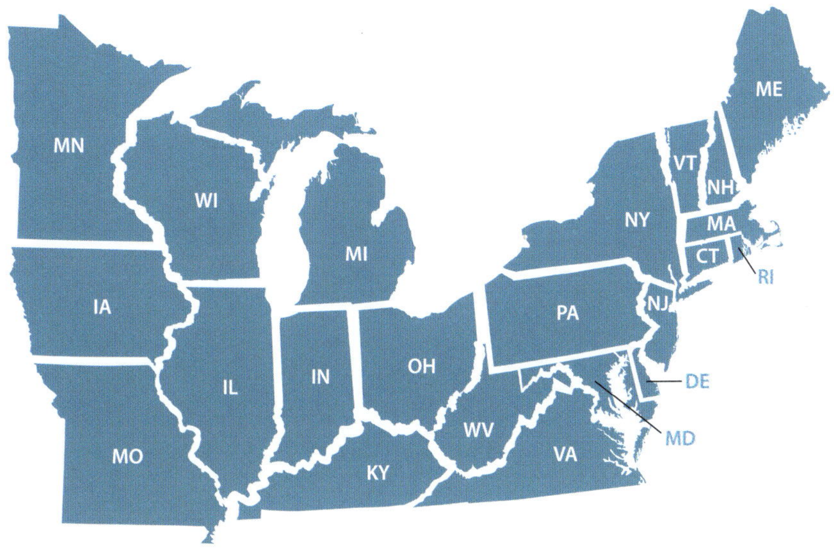

This book focuses on the northeastern quadrant of the United States, the area from Maine to Minnesota to Missouri to Virginia. This is one of the largest intact blocks of temperate forest in the world and has global significance for biodiversity and climate change mitigation. The forests of our region share defining ecological and social characteristics. Ecologically, the forest types of the region encompass temperate hardwood and conifer forests, such as northern hardwood, oak-hickory, and pine-oak-hemlock forests, as well as southernmost extensions of boreal ecosystems, including spruce-fir and aspen-birch. Writ large, these forests are in a state of recovery from previous periods of intensive land use, including extensive clearing and harvesting, that reset the ecological age of these forests to one much younger than what predominated prior to European settlement.

In addition, these forests exist in a region with a high density of nonindigenous insects and pathogens, resulting in a novel backdrop of uncertainty regarding future forest disturbances. This establishes an imperative to steward so that resilience is a key and common outcome, helping to ensure forests maintain their innate ability to adapt to and recover from environmental stresses.

Perhaps the greatest similarities across the region are the social characteristics of the ownership pattern. Our forests are owned mostly by private landowners, primarily individuals and families. Therefore, the decisions of private landowners regarding their approach to forest stewardship and the future ownership and use of their land are the main drivers of landscape change. The landscapes of this region also include federal, state, tribal, and municipal forests of varying sizes, and there is an increasing amount of land owned by conservation organizations. The relationship between the public and private conservation lands and private landowners is a topic of discussion throughout the book as we consider landscape perspectives.

PART ONE | Understanding Your Forest

“The gold of nature does not look like gold at the first glance. It must be smelted and refined in the mind of the observer.”
—John Burroughs, “The Art of Seeing Things”

CHAPTER 1

How Your Forest Works

TO THE UNTRAINED EYE, forests are simply a group of trees. Although it doesn't take training to appreciate forests for their beauty, improving your understanding of forests and the ways in which they work can enhance your connection to them and inform your decisions about them.

There are several ecological principles that can help you understand how your forest establishes, grows, and changes through time, naturally. Understanding these principles will provide you with insight into your options for stewardship. These principles are foundational to ecological forestry, serving as the template for the strategies discussed in the second part of the book. Learning about the ways in which your forest works can also help you develop a reciprocal relationship with your land so that you can help your forest thrive while also continuing to rely on it for everything you have come to appreciate about it and more.

A yellow birch grows on a rotting nurse log.

DISTURBANCE
Rethinking Forest Health

When landowners are asked about their goals for their forest, it is very common for them to include forest health as one of their top priorities. Landowners often equate healthy forests to those that are clean and orderly without any dying trees. However, it's important to flip that thinking right here at the start of this book. It's time to embrace messy forests and all the benefits they bring.

While excessive amounts of dead trees within a forest aren't healthy, we have to redefine "healthy forests" to recognize that dead standing trees (snags), downed dead logs, pockets of tree rot, and other structures that make forests more complex are critical and often missing parts of our forests. In fact, the very forest benefits we seek to provide, such as wildlife habitat and resilience, depend on them. The processes that lead to these important forest characteristics are known as *forest disturbances*.

COMPOSITION AND STRUCTURE

There are two important terms used throughout this book that are critical to understand right up front.

Species composition is the types and relative proportion of tree species within a forest.

Forest structure is the pattern in which the parts of the forest—trees, shrubs, understory plants, dead standing trees, logs—are arranged, both vertically and horizontally. Forests with simple forest structure have uniform tree diameters, one canopy layer, evenly spaced trees, few gaps in the canopy, and little standing dead and downed deadwood. One example of a forest with a very simple structure is a single-species plantation. Forests with complex structure have trees of different diameters, multiple canopy layers, irregular tree spacing, gaps in the canopy, and plentiful standing dead and downed trees. Old-growth forests are the model of complex forests.

The benefits forests provide—such as wildlife habitat, carbon storage, resilience, wood and other forest products, and water quality—are based on their species composition and forest structure. Forests with different species composition and structure will produce different benefits. Species composition and forest structure change through time, even without human intervention, and, therefore, the benefits forests provide change through time. Ecological forestry can be used to promote species composition and forest structure that provide desired benefits, by using natural patterns and processes as a guide to a forest's stewardship.

A disturbance is a discrete event that causes trees within the forest to be injured or die. Disturbances come in a variety of types, intensities, and frequencies. Disturbances in our region include wind events, ice storms, insects, disease, and fire. The most common disturbances involve wind (for example, thunderstorms, microbursts, tornadoes, hurricanes). Fire is more prevalent on drier sites and in the Lake States portion of the region. Insects and disease are also very common disturbances. The forests of our region are characterized by small (less than 3 acres), frequent disturbances. However, our forests do also experience periodic medium-size (3 to 15 acres) and large (more than 15 acres) disturbances.

The high frequency and small size of our common disturbances result in what's known as *patchy* forests—where some areas of the forests have been impacted by recent disturbance, some were impacted in the recent past, and others haven't been impacted by disturbance for many decades. These disturbances increase the structural complexity of our forests by diversifying tree sizes and ages,

A small-scale wind disturbance has blown down several trees in this forest, creating a gap in the canopy and increasing the amount of sunlight and diversity of conditions on the forest floor, including downed logs and mounds from uprooted trees.

In this medium-scale disturbance, a wind event has blown down most trees on several acres, creating near full-light conditions and leaving biological legacies, including downed logs and scattered surviving trees.

creating irregular spacing and canopy gaps, and contributing both dead standing and downed wood. Increasing structural complexity increases the amount of old forest habitat within the forest. Though less frequent, the medium- and large-scale disturbances play a critical role in creating young forest habitat for a suite of wildlife species that depend on these disturbed forests. So, while the immediate impact of forest disturbances results in dead trees, there is a bigger picture. Disturbances are part of the very fabric of our forests. Our forests and the full array of species within them have evolved over thousands of years with these various types, scales, and frequencies of natural disturbances.

Different disturbances affect different parts of the forest. Wind disturbances primarily impact trees with large crowns through leaf loss, branch and stem breakage, and whole trees being tipped over. A ground fire primarily impacts the forest floor. Insects often feed on a limited number of host species. The impact a disturbance has on a forest will therefore depend on the type of disturbance, its intensity and force, and the vulnerability the forest has to the disturbance. For example, a forest that has a lot of big trees in it is more vulnerable to wind events. A forest with a lot of oak species is vulnerable to the spongy moth, which feeds on the leaves of oak trees. The more vulnerable a forest is to a disturbance, the bigger its possible impact on the forest. Efforts to diversify forest composition and increase structural complexity help to spread out risk and avoid the loss through any single disturbance of a disproportionate amount of trees and the benefits they provide.

The Legacies of Disturbance

After a disturbance, it is common for our eyes to focus on what has broken or died. However, disturbances of all types and scales leave many *legacies* that help the forest sustain itself and the benefits it provides. There are almost always at least some individual trees and even groups of trees that survive to provide seed source and habitat. Seedlings and saplings, the seed bank within the soil, and roots that sprout also often survive in the understory, and all contribute to the regeneration of the forest after disturbance.

Another important legacy that disturbances create is deadwood. Standing dead trees, called snags, and downed logs and branches are important structural components of a forest. Snags are critical for wildlife habitat. As the tree begins to decay, it is host to many insects that help break down the wood. These insects serve as a food source for woodpeckers and other birds. The cavities made by woodpeckers hunting for the insects within are used by a number of other birds as well as small mammals. Once snags have fallen, the downed trees host an entirely different suite of animals, including salamanders. These logs store carbon and retain water in the forest. Downed logs also serve as "nurse logs," providing a moist, organic substrate for seeds to germinate. Species like eastern hemlock, yellow birch, red and white spruce, and northern white cedar have evolved to germinate on these nurse logs, making them very valuable for supporting forest regeneration. While it can be hard to see trees die, it's important to recognize that it's not the loss of a forest asset but a shift in benefits from those produced by live trees to the many essential benefits provided by deadwood. In this way, a tree spends part of its "life" living and the other part dead, but it remains a tree and a critical component of the forest the entire time.

Legacies serve as a bridge between the forest that was and the future forest, providing continuity from forest stage to forest stage and maintaining some benefits from the previous forest. The role and importance of legacies cannot be overstated. Many benefits of our forests that we value come from forest structures that take centuries to develop and are not easily or quickly replaced. For example, a 300-year-old sugar maple provides copious seed for regeneration, stores disproportionate amounts of carbon, and likely has cavities that are important habitat for a suite of species. It has taken 300 years to develop these wonderful benefits. It would take centuries for another tree to take its place as a large, old canopy tree. Since even very intense disturbances leave legacies, a forest stage always contains elements that are older than the last major disturbance. The importance and continuity of the legacies created through disturbance is a theme that we will revisit throughout this book.

Practicing ecological forestry means using the frequency, severity, and pattern of natural disturbances as a template for forest management, in both your woods and those of your neighbors. Doing so includes intentional efforts to leave legacies within the forest such as we would see after natural disturbances.

FIELDWORK

How Much Deadwood Is in Your Forest?

Rather than looking at live trees in your forest, note some of the results of disturbance by focusing on the deadwood. Start with the standing dead trees, or snags. Ideally, there are 5 to 10 snags per acre in your forest, each greater than 16 inches in diameter. How many can you see? The benefits they provide are related to their size. The bigger they are, the more wildlife (such as small mammals and owls) can use them. How big are they? Now look at the downed trees on the forest floor. Ideally, there would be 10 to 15 big (greater than 12 inches in diameter) downed trees per acre. How many do you see, and are they in different stages of decay? Finally, take a look up and see if you can find some gaps in the canopy where trees have died. Are the gaps the size of a single tree or are there gaps left by multiple trees?

REGENERATION

How Your Forest Renews Itself

Disturbance is only one part of the story. Fully functioning forests, those with all of their natural processes working properly and able to reach their full potential, will respond to a disturbance by regenerating native tree species using strategies these species have developed over thousands of years. For proof of the remarkable regenerative ability of fully functioning forests, we can simply consider the existence of our current forests, which have rebounded from intensive and widespread agriculture and timber harvesting that removed most of our forests over a century ago. Helping to sustain the inherent ability of our forests to renew themselves is critical to ecological forestry. This starts with a basic understanding of forest regeneration strategies. Successful regeneration often requires multiple regeneration pathways, which can be divided into two broad categories: sexual regeneration and asexual regeneration.

Sexual Regeneration

Your mother was right. It is about the birds and the bees. Sexual regeneration combines genetic material from a male and a female to produce offspring. In plants, including trees, this involves the pollination of flowers to produce fruits that contain seeds that will be dispersed in the forest. Like all living things, trees need to be a certain age before they are sexually mature and able to produce flowers. Trees that have attained sexual maturity have lived long enough to outcompete many of the other trees around them, surviving challenges such as insects, disease, browsing by deer, and drought. The strong survive. When these trees pass on their genetic material, they are often passing along the characteristics that help them be competitive and the raw genetic material that may be able to produce characteristics competitive in the future.

However, the challenges facing a forest change over time. New diseases emerge. Insects find new ways of getting around tree defenses. Climate conditions change. While the characteristics of the current sexually mature trees in the forest have made the trees competitive in the current environment, those same characteristics may or may not help the trees overcome future challenges. Trees adapt by diversifying their characteristics. Since sexual regeneration combines genetic material, it can produce unique individuals with some shared traits from the parents and some new traits. Trees with traits that make them competitive outcompete trees that lack those traits. It's survival of the fittest. One of the big advantages of sexual regeneration is trees' ability to adapt through time to new environments and challenges. In fact, trees are genetically very rich, providing a deep gene pool they can draw from to meet future challenges.

Seeds that fall into favorable moisture and light conditions can germinate and establish through sexual regeneration.

Rather than producing seeds in high abundance every year, trees typically produce large seed crops every few years. These large crops, often referred to as mast crops, help trees ensure some seeds escape consumption by seed eaters like insects and rodents, which have population sizes that are adapted to the more frequent low-seed production years. Tree seeds must land in an environment where they can successfully germinate and compete with other trees. Some seeds will find friendly environments with the proper light level and a reliable source of water. Other seeds end up in the seed bank of the soil, where they will stay either until they rot or until a disturbance causes an increase in sunlight, raising ground temperatures and stimulating the seed to germinate. Typically, smaller seeds that don't have a lot of stored energy are carried by wind. They thrive by landing on ground that has been disturbed or on decayed wood that provides a stable, moist environment. Several conifer species (trees with needles) associated with older forests, including red spruce and eastern hemlock, have small seeds that primarily establish on old, rotten logs or upturned soil created by a tree blowing over. The more deadwood in the forest, the greater the likelihood that these seeds will germinate successfully.

Larger seeds, like acorns and pine seeds, have enough stored energy to grow through leaf litter and even the turf of abandoned pasture to make contact with the mineral soil and a stable supply of moisture. This stored energy means that these larger-seeded species need not travel far from their parent to become viable seedlings.

Of course, as mentioned above, some seeds will be eaten by wildlife. Birds, such as blue jays, and small mammals, such as squirrels or chipmunks, play a very important role in spreading seeds, sometimes storing them in underground caches that can provide excellent conditions for germination. The movement of seeds by birds and small mammals (and historically humans!) is a great example of the interconnectedness of the forest. Healthy seed production helps maintain healthy wildlife populations and, in turn, healthy wildlife populations can lead to increased tree populations through the movement of seeds. In fact, some seeds actually benefit from the digestive process of birds, which can help break down the hard outer shell, increasing the seeds' germination rates when they are released from the bird.

Asexual Regeneration

If you've ever walked through your woods and wondered why some stumps have multiple stems coming out of them, the answer is stump sprouts, a form of asexual regeneration. Asexual regeneration is another pathway by which forests can regenerate themselves. As opposed to sexual regeneration's reliance on fertilization, asexual regeneration creates new individuals by sprouting new trees from parts of the parent tree. One of the most common examples is stump sprouts from hardwood species such as oak and maple.

After a deciduous tree (one that loses its leaves in the winter) is cut, the entire root system remains intact. The loss of the aboveground stem influences the flow of hormones in the tree, which, in combination with the energy stored in the root system, fuels the production of sprouts from buds on the stump. The cut tree is attempting to produce a new crown to make more energy and allow the tree to survive. Stump sprouts produce multistemmed trees in the forest. The diameter of all the stems of a multistemmed tree together is roughly the diameter of the original tree.

There are other examples of asexual regeneration. Trees may produce suckers, or shoots that grow from roots or the base of the tree. American beech, black locust, and quaking and bigtooth aspen produce prolific suckers. Trees that are adapted to riverbanks and floodplains, like willow and cottonwood, can also reproduce through fragmentation, when a part of the tree such as a branch gets carried downstream and then is buried on the bank and takes root. Species like spruce, fir, cedar, and tamarack can reproduce by layering, when lower branches touch the ground and take root, or tipping, when lateral branches of trees that have been tipped over turn up and begin a new individual. In each example, one part of the tree is the origin for the new trees.

The roots of a tree contain stored energy that fuel the growth of stump sprouts, a form of asexual regeneration.

Since new trees that are created through asexual regeneration don't involve fertilization, there is no new genetic material in the new trees. Instead, each new tree is a genetic copy of its parent. This lack of genetic diversity prevents the plant from developing new characteristics to help it adjust to new challenges, a distinct negative in a changing environment. However, asexual regeneration has the advantage of allowing trees to regenerate quickly. Instead of going through the long process of mature trees flowering and becoming fertilized, then developing, ripening, dispersing, and germinating seeds, individuals produced through asexual regeneration take advantage of the existing root system of the parent tree. Since the regeneration is genetically the same as the parent who already proved itself to be well adapted to the forest, the regeneration is typically well suited to the current environmental conditions. Being well adapted to the existing forest conditions helps a tree to quickly fill available gaps in the forest.

Multiple Pathways of Regeneration

Of course, after a disturbance, there isn't only sexual or asexual regeneration but rather a mix of the two. Tree species have evolved forms of each of these regeneration strategies. In the past, regeneration failure in our region was a concern only on extreme sites (for example, excessively dry soils), but with the spread of invasive exotic plants and increased deer and moose browsing, concern for successful regeneration has grown. The sexual and asexual regeneration strategies of diverse species give forests multiple regeneration pathways, increasing the likelihood that at least one of the species' strategies will prove itself successful in establishing the next generation of trees within the forest. These multiple pathways of regeneration are critical for forest resilience and ensuring the future of the forest and its benefits. Ideally, ecological forestry can help favor the regeneration of a diversity of specific desirable species by identifying trees to be retained in the forest and then removing an appropriate amount of trees in a way that will create the light and moisture conditions to favor those desirable species. However, the most important metric of success when regenerating a forest through harvesting trees is that new, native trees of some kind become established in a timely manner and in sufficient numbers to ensure the next generation of trees.

FIELDWORK

Where Is Your Forest Regenerating?

Focus on the next generation of your forest by looking for forest regeneration. Can you find seeds such as acorns, beechnuts, hickory nuts, and pine cones in the crowns of trees or on the forest floor? Do you see any seedlings in your forest? See if you can identify trees that have sprouted, such as sprouts on a deciduous tree stump. Are there parts of your forest with a lot of regeneration? Are there parts where there is no regeneration?

FOREST SUCCESSION

How Species Change over Time

Though trees are long-lived, that doesn't mean that the trees that regenerate immediately following a disturbance will be there forever. The species in your forest will change over time, even if you adopt a passive approach to the stewardship of your forest. The trees that initially regenerate after a disturbance will be those that are most competitive in the amount of light resulting from the small-, medium-, or large-scale disturbance.

All tree species can be categorized by the amount of shade they can tolerate to survive and grow competitively in the forest. This is known as the *shade tolerance* of the species.

Shade Tolerance

Some tree species have evolved to be most competitive through rapid growth in full sunlight. Their strategy is to establish, grow fast, and become the dominant trees within a young forest. We call these trees *shade intolerant* because they compete best using a lot of energy from the sun to grow quickly and, therefore, cannot thrive in the shade. Pioneer species such as aspen and paper birch are excellent examples of shade-intolerant trees,

SHADE TOLERANCE OF REGIONAL TREES		
	VERY INTOLERANT	
Aspen	Cottonwood	Pin cherry
Balsam poplar	Gray birch	Tamarack
Black locust	Jack pine	Willows
	INTOLERANT	
Bitternut hickory	Loblolly pine	Shortleaf pine
Black cherry	Paper birch	Sycamore
Black walnut	Pin oak	Virginia pine
Butternut	Pitch pine	White ash
Eastern red cedar	Red pine	Yellow poplar
	MID-TOLERANT/INTERMEDIATE	
American chestnut	Eastern white pine	Swamp white oak
American elm	Green ash	Sweet birch
Basswood	Hackberry	White oak
Black ash	Red oak	Yellow birch
Black oak	Shagbark hickory	
Bur oak	Silver maple	
	TOLERANT	
Black gum (tupelo)	Northern white cedar	Striped maple
Black spruce	Red maple	White spruce
Box elder	Red spruce	
	VERY TOLERANT	
American beech	Balsam fir	Ironwood
American holly	Eastern hemlock	Musclewood
Atlantic white cedar	Flowering dogwood	Sugar maple

doing well in areas of full sunlight, such as those resulting from a medium- to large-scale disturbance.

Other tree species are most competitive in partial sun and partial shade. These *shade mid-tolerant* or *intermediate* trees, such as red oak, white pine, and yellow birch, are highly competitive in light conditions that are roughly half sunlight and half shade. This type of light is often created when several nearby trees or a small group of trees dies, creating openings in the canopy that are referred to as gaps. Wind events, such as microbursts that touch down in parts of the forest, are a common way in which these gaps are created.

Still other tree species are *shade tolerant*. They grow very slowly and can therefore make do with low levels of light. Their strategy is to wait in the shady understory until an opening is made in the forest canopy. When an opening appears, such as after a windstorm or a responsibly planned timber harvest, they grow and take over that space. Species such as red spruce, hemlock, American beech, balsam fir, and sugar maple are all shade-tolerant species. Once shade-tolerant species come to dominate the main canopy of a forest, they often create shady conditions under which only they themselves or other shade-tolerant species can survive. Hemlock forests are an excellent example. Once established, hemlock often are the only trees able to regenerate in the understory, at least until there is a large disturbance that lets in a lot of light and tips the advantage back to species that are more competitive in higher light levels.

FIELDWORK

How Shade Tolerant Are Your Trees?

Look at the trees growing in the understory of your forest. Shade-tolerant trees typically have large crowns that come down low on the stem of the tree. Sometimes small shade-tolerant trees in the understory are as wide as they are tall. If you know some tree species or are comfortable using a tree guide, identify the species of trees in your main canopy, and use the table on page 29 to determine their shade tolerance.

The Successional Clock

Differences in the shade tolerance of tree species can help us predict changes in species composition over time. This process of change in vegetation composition on a piece of land over time is called succession. Immediately following a medium- or large-scale disturbance (like a hurricane or fire or an intensive timber harvest), shade-intolerant species are most competitive and are generally the ones to colonize the site. These fast-growing trees do not usually live long, however. After several decades they begin to die off, creating gaps in the canopy. Then, the mid-tolerant trees that are most competitive in these partial-sun/partial-shade conditions

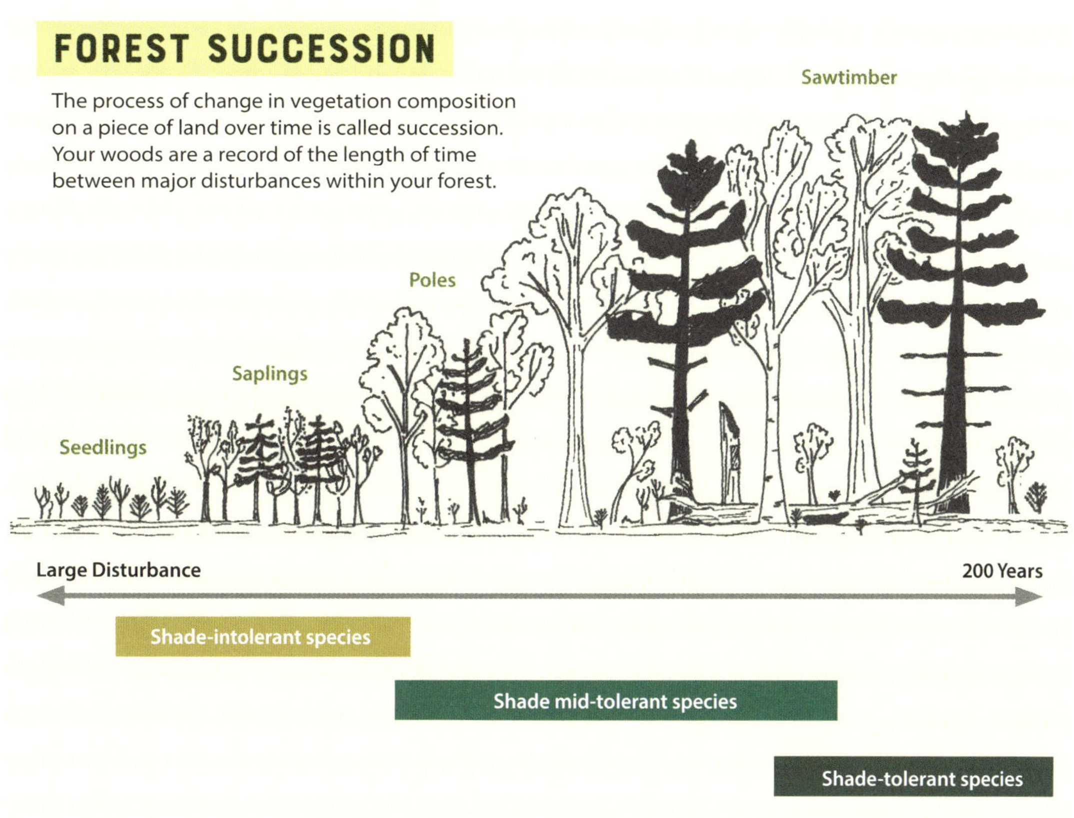

can grow and dominate the forest. They will sprout from the existing seed bank in the soil or be dispersed to the site from surrounding forest.

Over time, these shade mid-tolerant trees succumb to insects, disease, or weather events. Eventually, the shade-tolerant trees that have been "waiting" in the understory become the main canopy of the forest. In this way, your woods are a record of the length of time between major disturbances within your forest. Forests with a lot of shade-intolerant species in the main canopy of the forest indicate there was a significant disturbance within the last few decades, and forests dominated by shade-tolerant species often indicate that a lot of time has passed since the last significant disturbance.

The usual understanding of succession is as a one-way, linear process: Forests progress from shade-intolerant to mid-tolerant and finally to tolerant species. However, nature is rarely linear. It is more accurate to think of succession as a clock. The hands of the clock represent the amount of time since the major disturbance. Forests will continue their path around the clock until they are disrupted by another disturbance, setting the clock of succession back. If it's a large disturbance, like a hurricane, impacting most of the trees in the forest, the successional clock can be set back to the beginning (time zero), favoring shade-intolerant species. If the disturbance is less intense, affecting only a portion of the trees in the forest, it is likely that it will favor shade mid-tolerant trees (time 50).

THE SUCCESSIONAL CLOCK

The hands of the successional clock represent the amount of time since the last major disturbance. When a new disturbance occurs, the clock will be set back—or sometimes even forward.

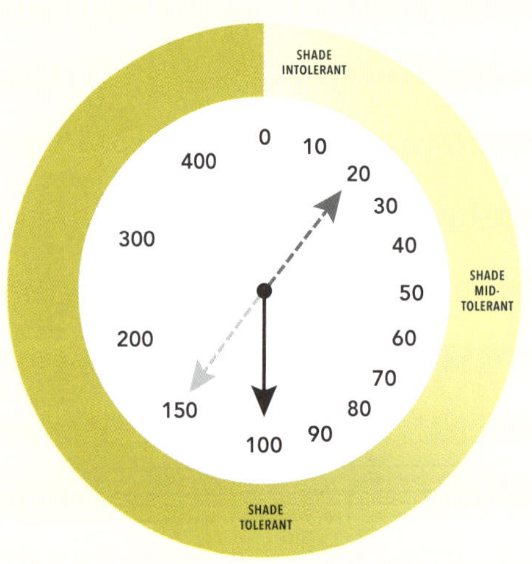

Importantly, disturbances don't always move the hand of the successional clock backward. There are times when disturbances actually speed up the successional clock. An example of this is a forest that is dominated by shade mid-tolerant species, such as red oak and white pine, in the overstory with an understory of shade-tolerant eastern hemlock. A disturbance that causes mortality to the overstory, releasing the understory, would decrease the time it takes the overstory to become dominated by shade-tolerant hemlock. This is represented by the dotted arrows in the figure below left.

The successional clock underscores that your forest is a highly dynamic ecosystem. Most days your forest is moving forward in succession. However, disturbance is at work in your forest. Disturbance can move the hands of the successional clock back, favoring early succession species, or speed up the clock, favoring more tolerant tree species. Day after day, year after year, decade after decade, century after century, your forest changes. Succession has no end point.

Since each species has its own unique benefits, each stage of succession brings with it its own combination of benefits. When evaluating your options, it is important to consider which stage of forest succession is providing the species composition associated with the benefits you seek from your forest. You can use ecological forestry practices to move the hands of the successional clock backward or forward to help the forest (or parts of it) achieve the desired successional state.

The preforest stage is temporary and typically hosts the greatest amount of biodiversity.

FOREST DEVELOPMENTAL STAGES

How Structure Changes over Time

While the process of forest succession helps us better understand the way in which forests change in species composition over time, it's important to also understand the way in which forest structure changes through time. Like succession, changes in forest structure are influenced by disturbance. Forest development—tree growth and competition with one another—is also an important factor, helping to move forests to greater structural complexity. The following descriptions of the stages of forest development start just after a large disturbance, when the hands of the successional clock have been turned back to time zero, and continue through the old forest stage.

Stage 1: Preforest

After a major disturbance, the forest is not actually dominated by trees but instead by plants such as grasses, forbs, and shrubs, which temporarily take over the site. Biological legacies remain, including scattered live trees, roots, the seed bank, and often plenty of deadwood, both standing and downed. This is the *preforest* stage of forest development. Despite our love of forests and interest in seeing them regenerate as quickly as possible, this temporary pre-tree stage plays a valuable role. For example, there are a number of wildlife species, including those that are in decline and even threatened, that depend on this specific preforest stage. These include a number of bird species, such as the chestnut-sided and golden-winged warblers, and invertebrates such as butterflies and bees. In fact, of all forest stages, this one has the highest amount of biological diversity (commonly referred to as *biodiversity*, the variety of all living organisms), due to its many habitat niches and food sources. However, the preforest stage is very ephemeral, often lasting no more than 15 years before transitioning to the next stage.

Stage 2: Young Forest

The second stage of forest development, the young forest stage, brings with it significant change as the site becomes dominated by trees again. As trees establish and begin to form a canopy, they significantly reduce the amount of light reaching the understory. As a result, the rich herbaceous layer of grasses and shrubs that defined the preforest stage can no longer persist. The young forest is typically made up of shade-intolerant and mid-tolerant tree species, since these species have evolved to take advantage of full sunlight. During this stage of forest development, the forest has a highly simple structure, consisting of thousands of trees per acre of the same size and age, evenly spaced, and with one canopy layer. The result is a lack of

This young northern hardwood forest is approximately 20 to 30 years post-harvest.

habitat diversity, making this stage of forest development the one with the least amount of biodiversity.

There is severe competition among the trees as they strive to grow fast and gain growing space, water, and nutrients. The saplings are absorbing carbon from the atmosphere as a raw material for photosynthesis, shooting up like teenagers going through a growth spurt. This competition and growth rate result in the highest rate of carbon sequestration of any stage in a forest. The winners that are best adapted to the site and conditions begin outcompeting the trees around them, gaining more and more space in the main canopy. The losers begin falling behind in the race to stay in the main canopy, eventually ending up in a subdominant crown position and ultimately dying from a lack of resources—primarily light, but also water and mineral nutrients.

This stage has been described as one of "stem exclusion," since the full occupancy of the forest at first prevents any new trees from establishing and, later, results in some trees dying. The competition among these young trees and the structure it creates lead to the next stage of forest development.

Stage 3: Mature Forest

Over several decades, the forest continues to grow, develop, and change, leading to the mature forest stage. The main canopy begins to differentiate, with some species, such as red oak and white pine, expressing their growth habits and taking on upper, dominant

This mature forest is 80 to 100 years old, with shade mid-tolerant trees in the overstory and shade-tolerant trees in the understory.

positions within the forest canopy, while other species, such as black birch and red maple, assume lower, subdominant crown positions. The crown differentiation adds canopy layers within the forest, increasing structural complexity. As trees succumb to disturbances such as insects, disease, and wind events, gaps form within the canopy. These gaps begin to regenerate through the pathways we've discussed, with the dead (standing and downed) trees adding to structural complexity. With the increased structural complexity of this developmental stage comes increased biodiversity and carbon storage benefits.

Stage 4: Old Forest

An old-growth forest is one that has never been harvested or been cleared for agriculture. This is an exceptionally rare condition in our region. Once a forest has been taken out of old growth, it can't be returned to old growth. However, if enough time goes by, a mature forest can accumulate enough characteristics that make it comparable to old-growth forest. At this point, the forest enters the *old forest* stage of forest development.

In an old forest, there has been a long time since the last major disturbance, so centuries of competition have played out, leaving a handful of very large trees in the forest. Meanwhile, small disturbances over this long period of time have created a patchy forest with a large amount of structural complexity—trees of different sizes and ages, dead standing trees, downed deadwood in various states of decay. This offers a niche-rich environment for wildlife, making the old forest stage the second most biologically diverse of the forest stages, behind the preforest stage. This is also the stage of forest development with the highest amount of carbon storage.

The forest will continue at this stage of development, being subjected to small disturbances and increasing in complexity, until a large disturbance eventually occurs, turning the clock back to an earlier stage of succession and starting the process again.

FOREST STAGES ON THE SUCCESSIONAL CLOCK

Forest stages can be placed on the successional clock as well. Older forest stages correspond to greater shade tolerance.

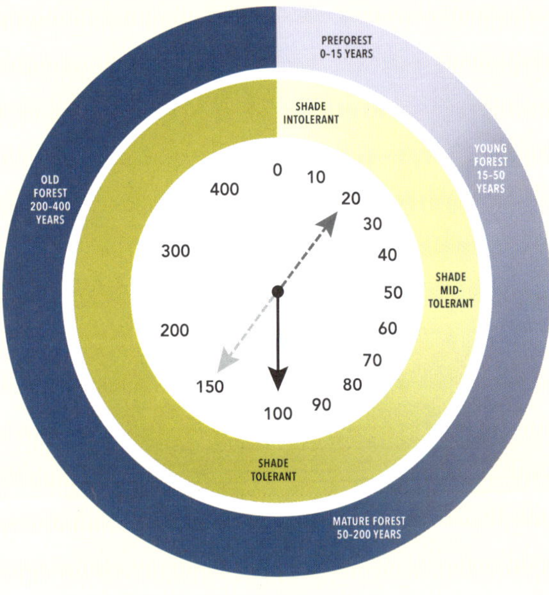

This old forest is approaching 200 years old, with large, old trees, young growing trees in a gap caused by disturbance, and ample dead standing and downed trees.

A Shifting Mosaic

Defining stages of forest development helps us better understand the change in forest structure over time, but in nature the lines between the stages aren't so neatly defined. Instead, there is a gradient between these stages within a forest. Also, while it may be easiest to envision the stages of forest development in a linear fashion, where a forest goes through each stage one after the other, it's more accurate to think about different stages happening around the forest simultaneously.

Disturbances happen in different parts of the forest at different times, in different ways, and with different severity, leading to a mix of tree species, sizes, ages, and spatial arrangements. Each year in any given forest, there are trees that die and leave gaps in the canopy, resulting in preforest patches. As these patches within the landscape develop into the young forest stage, disturbances occurring in other parts of the landscape create new preforest patches. This movement of preforest stage across the landscape is known as a shifting mosaic. So, although a forest may be one hundred years old, it also contains patches that are younger and in different stages of development. This dynamic process helps create the patchy, complex structure of mature and old forests, an important characteristic that ecological forestry seeks to mimic.

FIELDWORK

What Is the Developmental Stage of Your Forest?

When making this determination, focus on the structure of your forest. Look for key characteristics of each of the developmental stages:

Preforest: no/few trees, dominated by grasses and shrubs

Young forest: many small-diameter (less than 10 inches) trees, no understory

Mature forest: multiple canopy layers, larger trees, understory

Old forest: some very large trees (more than 20 inches in diameter), patchy forest, a lot of deadwood

Remember, your forest may not neatly fit into one category, as it may be transitioning from one stage to the next. Also, different parts of your forest may be in different stages, depending on the disturbance history.

Each stage of forest development has its unique benefits. When evaluating your options for practicing ecological forestry, it is important to consider which forest developmental stage or combination of stages is providing the species composition and forest structure that will help you achieve your goals for the forest. Are you interested in supporting wildlife and biodiversity in the preforest stage? Do you want to regenerate shade mid-tolerant species? Are you interested in maximizing carbon storage? This is how you use ecological forestry to move the hands of the successional clock backward or forward to achieve the desired forest developmental stage(s).

THE LANDSCAPE CONTEXT

Making an Informed Decision Within the Bigger Picture

No forest, no matter how large, exists in isolation. Instead, it is connected to the landscape all around it, both influencing it and being influenced by it. Likewise, landowners are connected to others within their social landscape. There are myriad neighbors, friends, and professionals with whom to share knowledge and experience. To make an informed decision about the stage(s) of forest development that will best meet your goals, it's critical to step back and take a look at the big picture. How does your forest fit into the wider ecological landscape? How do you fit into the social landscape?

The Ecological Landscape

Our region is a quilt of forest ownerships. The majority are owned by families and individuals and include properties ranging in size from a few acres to a few hundred acres. In addition to family forest owners, public agencies (town, county, state, and federal), conservation organizations (land trusts), and tribes also own forests across the landscape, often in property sizes far larger than family forests. Similarly, large private properties may also be owned by corporate forest owners that manage lands primarily for financial returns. Together, these various ownerships all fit together in the landscape like puzzle pieces, each with their own unique characteristics and roles.

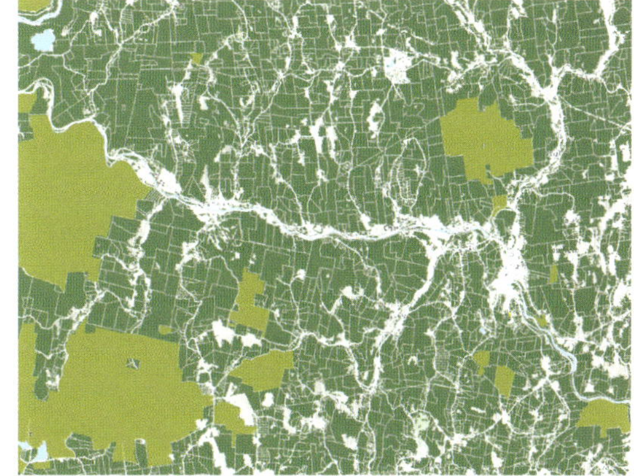

The properties of private landowners dominate the landscapes of our region and fit together like puzzle pieces, as in this parcel map. The parcels of private landowners are shown in dark green. Public ownerships (olive green) often serve as anchor points within the landscape.

The forest benefits that we seek to sustain often require more land area than that of one individual property. For example, for those interested in wildlife, the home range of a black bear is 15 square miles (almost 10,000 acres), and a turkey needs 0.5 to 3 square miles (400 to 2,000 acres) to meet its needs. If you're interested in water quality, watersheds are tens of thousands or even hundreds of thousands of acres. Even recreation, such as hiking and snowmobiling, is more fun on trails that extend beyond the boundaries of a single property. Challenges, too, exist on larger scales. Invasive plants and insects and wildlife habitat are most effectively addressed by multiple landowners at scales larger than one property.

While you have decision-making agency over only your own property, your land is clearly connected to the surrounding landscape. As part of this quilt of connected pieces, your forest and the choices you make about it affect not only the personal benefits you receive from them but also those received by your neighbors and the community. Similarly, the way in which you approach challenges, such as invasive plants, affects those around you, and likewise the decisions others make in the landscape impact you. Every decision and nondecision influences the forest and the benefits the landscape produces.

Components of the Landscape

The first step in evaluating the landscape is to understand its components and their configuration. Every landscape is unique. Your landscape may include land uses such as agriculture, residential housing, and commercial development as well as water features such as streams, rivers, lakes, ponds, and wetlands. Some landscapes are heavily forested with occasional houses and small farms breaking up the forest. Other landscapes are suburban with primarily residential neighborhoods and patches of forest separated by houses and roads. And, of course, there is everything in between.

Some parts of the forested landscape hold unique ecological and social value. These may include water features (streams, wetlands, and vernal pools), endangered species habitat, and unique site characteristics that are not typical in the larger landscape (for example, very wet, very dry, high elevation, or low elevation), talus slopes, and very old forest. These areas contribute to the overall function and integrity of the landscape, including the biodiversity it supports, and deserve special conservation consideration.

In addition to its ecological resources, our region has a long history of human and forest interaction. For the thousands of years before colonial times, the forests of our region were home for many Indigenous peoples. These cultures have their own unique, culturally important areas of forests, such as the locations of ceremonies or historic gatherings and events. In addition, Indigenous cultures have a close relationship with certain plant and tree species, such as black/brown ash and northern white cedar, that provide culturally important materials for food, medicine, and spiritual use. Finally, there are significant cultural resources from the colonial era, such as foundations, stone walls, mill sites, and logging camps, as well as sites of important events. As a result, there may be places within the landscape that carry special cultural value in need of protection.

Building Forest Connections

Generally speaking, the larger and more connected a block of forest, the better that forest is able to function properly. Protecting forests from conversion to other land uses to ensure a critical mass of forest in the landscape is an essential first step in achieving this condition. Larger blocks of forest preserve whole habitats for plants and animals and reduce the *edge effect* that other land uses may have on forests: Any land use that is not forest creates an edge, or interface, between different land uses. Sometimes these edges are hard, like a forest next to someone's backyard where there is a distinct line between the forest and the lawn. Other times the edge is soft, like a forest next to an abandoned field filled with shrubs. These edges can alter forests by introducing invasive species and even pets (such as cats) and livestock into the forest, inhibiting forest function. Larger patches of forests have more core area that is untouched by the edge effect, allowing the forest to function without hindrance.

Edges may also stop trees, plants, and animals from moving across the landscape. As discussed in the regeneration section (page 24), each tree species has strategies it uses to regenerate itself. If the trees aren't able to cast seed into favorable environments, or if the wildlife populations that trees depend on to move seed are reduced, the trees' chances for regeneration are also greatly reduced. This becomes particularly important as forests adapt to changes in climate. The ability to migrate across the landscape will be critical to tree species' survival by allowing them, when conditions where they currently exist change and become less favorable, to spread into more favorable environments. Of course, trees are not the only species that need to move through the landscape. Animals also must do this for their survival and reproduction. Moving through connected areas of habitat increases their likelihood of successfully meeting their needs.

In short, we want a lot of forest on the landscape and, ideally, we want that forest to be in big, connected blocks. Your decisions can help keep forests as forests and keep them connected. Engaging in conservation-based estate planning (see Chapter 12) will help ensure your forest remains forest in the future. Focusing on developing complex structure within your forest will help create

Every landscape is made up of a unique arrangement of components such as forest, agriculture, residential development, and water.

stepping stones—patches of similar habitat—for species to move through the landscape. Your decisions about the forest stage or combination of forest stages that will help you meet your goals can also encourage movement of species through the landscape by providing continuous habitat.

Considering Landscape Targets

So, what's an ideal mix of forest developmental stages across a landscape, and how can your land contribute? To answer this question, we can look at the past to get a sense of the forest stages that were created through natural disturbance and Indigenous cultural practices, like burning, to estimate the conditions that wildlife species would have evolved with over thousands of years.

In general, upward of 90 percent of our region was covered in what we now call old-growth forest, with the other 10 percent being in a preforest condition that shifted around the landscape and was caused by disturbances such as wind events, naturally occurring fire, Indigenous burning, and beaver flooding. This proportion varied across the region and forest types, with 20 to 55 percent of fire-dependent jack pine and white pine–red pine ecosystems in the Lake States historically supporting old growth. However, using the past as a guide, it is safe to say that in most parts of our region, the majority, and in many cases the vast majority, of our forests should be in the mature or old forest stage, with a small but important percentage in the preforest stage.

Given that these are the approximate proportions of forest stages that our forest species evolved with, working toward these targets within your landscape will help ensure the suite of habitats necessary to sustain native biodiversity. Determining the ways in which your forest can help create or maintain the ideal mix of forest stages in the landscape while achieving your own goals is an important consideration when evaluating your options for the stewardship of your forest.

FIELDWORK

What Is the Landscape Context of Your Forest?

Look around your landscape. How much of it is forested? Is your forest connected to other forest? What other types of land uses are there in the landscape? Are there water resources or special habitats in your landscape? Finding a map of your landscape can be very helpful in evaluating the context of your forest. Many states have online mapping tools. Your forester or local land trust can also provide assistance. Refer to Resources (page 264) for contact information in your state.

THE ROLE OF THE SOCIAL LANDSCAPE

While your land plays an important role in the ecological functioning of the landscape, you and your neighbors play an important role in the social functioning of your landscape. The landscape around your land likely has hundreds or even thousands of landowners. Each of these landowners is making decisions about their land as choices arise in their lives. Sometimes decisions are triggered from outside the house, such as a knock on the door with an offer of money for timber or house lots. Other times the decisions are triggered from within the house, from the circumstances of one's life, such as retirement, health issues of a loved one, or a child going off to college.

The combined public benefits that all family forest owners provide depend on the decisions of hundreds and thousands of landowners. It gets complicated! The good news is that with so many landowners across the landscape, there are wonderful opportunities to learn from one another. In other words, the stewardship of your forest isn't just about your relationship with the forest; it's also about your relationship with other landowners.

When making decisions about your forest, it's natural to turn to peers, friends, and neighbors for input. There are very good reasons for this: Peers may have experience with the decisions you are making and can share what they learned from their processes and outcomes. What went well? What would they change? Peers have local knowledge, and they give it to you straight, without the use of jargon or government program acronyms. Peers know local professionals, neighborhood dynamics, and possible pitfalls, and they don't have a financial interest in your decision. (When you walk into a car dealership and the salesperson tells you about the great deal they have for you, it's natural to wonder if it is a very good deal or if the person stands to gain a great commission if they sell you that car.) Peers, especially your friends, often share the same values that you have. That's why you're friends with them. Peers are also available for time-sensitive consultation. Decisions about one's land can arise for any number of reasons, many of which are unplanned. There is rarely a landowner workshop or webinar coinciding with your decision. It is no wonder that landowners often turn to friends to help inform their decisions.

You are surrounded by other landowners with knowledge and experience. And this isn't a one-way street. The more experience you have with the care and stewardship of your land, the more information and experience you have to share with others. Of course, accessing the social network of your landscape means either having a relationship with friends and neighbors who have knowledge and experience or, at least, having the opportunity to speak with them.

Recipes for Connection

While much of this book focuses on the ingredients of ecological forestry and how to improve our relationship with forests, our relationships with other people are equally important. The more relationships people have with one another, the more information moves through the social network of landowners in a landscape, which increases opportunities for landowners to find the information they need to make an informed decision.

The "recipe" for increasing social relationships is all about getting people together. Neighborhood potlucks can change the world! So how do you increase the social capital within your landscape? Try hosting a walk on your land, a neighborhood potluck, a bonfire, a pie-and-coffee social, an invasive plant–pulling party, or a mushroom log inoculation. You could consider sharing equipment (like a log splitter or brush hog), connecting trails, or organizing a deer hunt. When you get people together you don't have to worry about having a formal agenda. Take the pressure off. Simply gathering around the topic of land will inevitably lead to good things. It would not be unusual for the conversation to start with comments about the crazy weather you've been having, then move to what the kids are up to, and eventually to the land. It's also important to recognize that some landowners may be more interested in connecting than others. Focus on those who see the value of connecting and sharing information.

After all, it's not a matter of *if* you will face a decision about your land; it's only a matter of *when*. The same goes for your neighbors. Building neighbor relationships may also provide opportunities to help one another reach mutually beneficial goals, such as building trails, controlling invasives, and implementing habitat management across boundaries.

Also, keep in mind that in this internet age, the word "neighbor" has come to have a very fluid definition. Landowners are part of a community of place and communities of interest. These days, your "neighbors" may live across town or across the country. Geographic proximity may be less important than the shared experience of being a forest owner or sharing a specific interest, such as deer hunting or foraging for fungi. Regardless of where you find neighbors, we are all better off when we help one another out by sharing our knowledge and

WOMEN OWNING WOODLANDS NETWORK

Getting together with people you can relate to and are comfortable with can help facilitate relationship building and information exchange. The Women Owning Woodlands Network (WOW) is a national effort to address barriers that women woodland owners face in stewarding their land. WOW brings female landowners together, providing safe spaces to learn and to share their experiences. To find the nearest WOW chapter, visit womenowningwoodlands.org.

experience. All landowners have something to offer.

Neighbors are not the only important social relationship for your land. As a family forest owner, it is highly likely that you own the land with someone else (such as a spouse) and that your decisions about the property, particularly those about its future ownership and use, include other members of your family, such as children. Family dynamics can be complex. There can be differences in opinion among family members who may have varying relationships to the land, different financial circumstances, and varying geographic proximity to the land. When it comes time to making decisions as a family, creating an inclusive and welcoming atmosphere is likely to facilitate productive conversations. A family dinner or walk in the forest can be a great opportunity to talk about your family's feelings about the land and what they would like to see happen to it.

FIELDWORK

How Can You Share Information?

Think about the types of information you may need to make informed decisions about the stewardship of your forest. What information could another landowner potentially provide you with? What questions would you ask them? What information or experience can you share with someone else making a decision?

HUMILITY
The Final Ecological Principle

Each forest is an ecosystem—a wonderfully complex, dynamic, and connected community. Despite all that is known about forests, it's vitally important to approach all forest management with a sense of humility and openness to different ways of knowing. Though the last couple of decades have seen an explosion of Western science research to understand the ways forests work, we most certainly do not know all there is to know and are only beginning to appreciate other worldviews, including Indigenous perspectives. Forests are also constantly evolving and changing in ways that we've never experienced before. Simply put, we cannot assume we have all the answers.

Practicing ecological forestry means questioning our assumptions, being keen observers, and adapting our strategies to new information. However, we cannot let perfection be the enemy of the good. We must move forward with deepening our relationship with forests by engaging with them, but we must do so cautiously and carefully, working well within the limits of forests and stressing restoration and resilience. Doing so will encourage forests to function fully, express themselves completely, and preserve their integrity and their innate ability to self-maintain. Through this practice, we, too, will be changed by the experience and our growing understanding.

PULLING IT ALL TOGETHER

A Walk in the Woods

Let's take a walk and see how the principles of forest ecology work together.

Life has been busy. It's been a while since you've been able to walk your woods. Some things have changed. Many of the trees are the same, though they have grown larger. You notice the main canopy includes some shade-tolerant sugar maple as well as shade mid-tolerant white ash and yellow birch, most likely established when gaps were created in the forest canopy decades ago. Underneath are shade-tolerant sugar maple, hemlock, and American beech seedlings and saplings waiting for their opportunity to grow into the main canopy.

As you walk down the trail, you notice a small group of trees covering two acres that were hit by the microburst a couple of years back. Of course, not all the trees were killed. There are a number of biological legacies. A large sugar maple remains standing in the gap, along with a number of sugar maple saplings. There is a snag from a treetop that snapped off, the bark sloughed off with woodpecker holes up and down it. Other trees killed by the storm have fallen to the ground, adding to the downed dead logs in the forest. Yellow birch and hemlock seedlings grow on the decaying log. Salamanders find cool, shady conditions under it.

In the two acres of sunlight created by the open canopy of the trees that have died, the shade-tolerant sugar maple saplings that were once in the understory are now getting their

Taking time to regularly walk through your forest and watch it change is great way to get to know it.

chance to reach for the sunlight. The suckers of a nearby aspen are also growing in the sunlight. There are even seedlings from shade mid-tolerant species, white pine and red oak, from the surrounding forest that have seeded into the gap and germinated, thanks to the increased light levels. Each of these regeneration pathways is helping ensure the future of the forest. As you walk, your ear recognizes the calls of several new bird species you haven't heard in the past, and you realize that they are using the multiple canopy layers within the gap.

It's not just the forest that has changed. As you come out of the forest you realize that the landscape around the forest has changed as well, bringing new challenges. You notice Japanese barberry from the nearby pasture beginning to establish in the understory of your forest. There are more houses on the road, reducing the forest cover and connectivity and increasing the amount of edge. You remember that your neighbor recently passed away, and house lots were sold off to settle the estate. You are reminded that you keep meaning to talk to your spouse and children about the future of the land. Who will own it when you're gone? How will the land be used? One of your friends just worked with a land trust to make sure her land isn't turned into house lots. You make a mental note to give her a call and find out more about her experience.

You leave the forest relaxed. Your mind is clear, and you even notice your heart rate is slower. Being in the forest helps you appreciate its importance to you, your family, and the community. You ask yourself why it's been so long since you walked in your forest. Out of sight, out of mind, you guess. You make a date with yourself to take another walk this weekend. Practice makes perfect.

"If you don't know your past, you don't know your future."

—Ziggy Marley

CHAPTER 2

Forests of the Past

BEFORE COLONIZATION, about 90 percent of our region was forest or savannah. Today, it remains one of the most densely forested regions in the US, with forests covering roughly a third to almost 85 percent of various land areas. These forests are the vegetative expression of our ample precipitation and fertile soils. In the warmer, southern portions of our region, the forests are predominantly mixed oak-hickory and pine-oak, while to the north we have northern hardwood; spruce-fir; and jack, red, and white pine forests. Savannahs largely dominated by oaks also become more common as you move to drier portions of the Midwestern states covered by this book.

For thousands of years, forests in these areas established, grew, and changed, driven by the ecological principles discussed in Chapter 1. Through this evolutionary process, the best-adapted species sorted themselves into each landscape according to factors such as temperature, precipitation, and the length of the growing season. But our forests have been shaped by more than ecological processes. As humans evolved within and among forests, forests were continually shaping humans—and, to varying degrees, humans were shaping forests.

A stone wall runs through a forest, serving as evidence of the intensive agricultural land use of our region in the colonial era.

INDIGENOUS RELATIONSHIP TO FORESTS

Descriptions of the forests of our past too often start with European settlers. Before settlers arrived, however, there were thousands of years of relationship between forests and the rich cultures of Indigenous peoples of our region. These peoples had a direct and close relationship with forests, which included harvesting forest plants, trees, and animals for various uses, including food, medicines, ceremonies, building materials, fuel, and canoes—a relationship that continues today.

Forests were and are intimately tied to Indigenous culture. The creation story of the Wabanaki people begins with a brown ash tree. The Haudenosaunee (Iroquois) use white pine in their wampum and oral history to help demonstrate their confederacy's deep roots, stature, and strength. These and countless other examples from tribal nations across the region covered by this book highlight the belief that, rather than viewing forests as an impediment to be conquered, as was the prevailing attitude of European settlers, many Indigenous cultures view all living beings as related and each an equal and important part of creation. This fosters a relationship of reciprocity with the forest. There is acknowledgment of the many gifts the forests provide and, in return, a sense of responsibility to reciprocate these gifts by caring for one's relatives: the trees, animals, land, and water. It is clearly understood that there is a shared fate among all beings and a decision time frame that extends for generations. Rather than thinking of the implications in months or years, the implications of decisions are judged based on generations and millennia.

The relationship of Indigenous peoples with the land should not imply that they didn't steward forests. On the contrary, part of the reciprocal relationship is active engagement with forests. Indigenous cultures manipulated forest and prairie ecosystems to meet their needs. This included harvesting plants and trees and also using fire to encourage production of food (such as berries and deer habitat) and to make it easier to see enemies at a distance. Indigenous peoples also assisted plant migration. While forests were primarily shaped by natural disturbances and processes, forests and Indigenous peoples evolved together, each influencing the other.

Removal of Indigenous Peoples

This balance and relationship of reciprocity remained in place until European settlement and colonization. Explorers were given the right to claim lands for their country if they were not inhabited by Christians. The result was the violent seizing of land and the forced removal of Indigenous peoples from their homelands. Though uncomfortable for those of us of European descent, a conversation about practicing ecological forestry for landowners cannot ignore the reality that the land we seek to care for and steward holds this history.

Institutional policies of the US federal government, some of which continue to this day, were designed to deny Indigenous peoples their rights and to eliminate the people themselves. The impact on Indigenous groups has been profound, decimating some cultures and driving most to the edge of extinction. However, many Indigenous peoples have been remarkably resilient and, despite the overwhelming past and present challenges, have persisted. Encouraging efforts to sustain Indigenous peoples and their lifeways are being implemented in many places. However, given the intimate relationship between Indigenous peoples and their homeland and waters, to fully preserve and maintain these cultures, they need more access to land.

Ecological forestry is about restoring all the parts and functions of a forest—ecological, cultural, and social. Doing so can also help restore Indigenous cultures by providing the full suite of forest components Indigenous peoples rely on for traditional food, medicine, and cultural activities. Each component plays a role in the lifeways of Indigenous peoples. Cultural Respect Easements (see page 256) are one tool that can help facilitate Indigenous access to the land.

FIELDWORK

What Is the Indigenous History of Your Land?

Your land is the homeland of Indigenous people that lived there before they were forcibly removed. Determine which tribe or tribes call the area where you live their homeland by visiting https://native-land.ca.

COLONIZATION AND CLEARING

European settlers brought with them a different relationship between humans and the forest than that of Indigenous peoples, and our forests are still recovering from the effects. To Europeans, the forest seemed both an inexhaustible resource and an impediment. From the start of colonization, there was a fervent and sustained effort to tame the wilderness and make it "productive." To the Europeans, this meant converting much of the forest into agricultural land, as areas of wild forest were considered "wasted" land and viewed as dangerous—or even inhabited by the devil.

The first wave of European settlers in the US, including in what became New England, went to work establishing subsistence farms in the rocky soils to meet the needs of their families. Most of the area was dedicated to pasture with a much smaller proportion used for crop production. As the number of European settlers increased, more forest was cleared. Subsistence farming turned into market farming as population centers grew. Agriculture peaked in New England around the mid-1800s, at which time the forest cover across much of this region was reduced to approximately 30 percent of the land area. The forest that was left was often in patches of woodlots that provided firewood and building materials for farms. As settlement expanded west to the Ohio River Valley and beyond, farmers recognized that the deep, rich soils were easier to farm than the rocky soils of New England.

Transportation networks of roads and rails made it possible to move agricultural goods east, further reducing incentive to farm on marginal lands. In addition, there were increased employment opportunities in urban centers as the industrial revolution took hold. Farms were abandoned, and forests began to recover in the Northeast. Meanwhile, the pattern of forest clearing for agriculture moved like a wave into the Midwest as cities and towns sprang up, built primarily out of the forests that existed within a river log drive or railway's reach away. Large areas of the upper Midwest were deforested in a very short period of time. Areas that weren't cleared for agriculture served as the source of raw materials to support growing communities and industries, like commercial lumber and pulp and paper production, resulting in large areas of "cut over" forests. Our forests did not evolve under disturbances of this scale and completeness.

The Scale of Human Disturbance

One of the important characteristics of natural disturbances is that they very rarely result in the death of all the trees in the forest, and many biological legacies remain to provide continuity through the next stage of the forest. But with colonial-era forest clearing,

This diorama depicts the peak of agricultural clearing of forests in Massachusetts in the mid-1800s.

A historical and a contemporary photo of the same property show the increase in forest cover in our region since its agricultural past, demonstrating why our forests are ecologically young.

virtually all the trees in areas to be used for agriculture were cleared. Standing and downed deadwood, as well as leaf litter, were all removed. Some of the land was plowed, mixing soil layers and breaking down organic matter. The environmental diversity of the sites plummeted. Natural disturbances are never so complete.

The use of remaining forests as woodlots and sources of raw materials was also different from natural disturbances. In many woodlots, the first harvests were often of low severity, focusing exclusively on the highest-value individuals and species (for example, white pine, red spruce, oak) and leaving behind smaller and less economically valuable trees. The development of new uses, such as pulp and industrial charcoal, meant that eventually all trees gained market value, resulting in more severe and extensive clearcutting. These waves of cutting, sometimes followed by fires associated with the railroads, resulted in a general homogeneity in the age, structure, and composition of forests across large areas, a stark contrast to the mosaic of ecological conditions created within and across forests by natural disturbances.

Given the scale of these intensive impacts, only a tiny fraction (less than 1 percent) of forests in our region have been untouched by the agriculture and logging activity of the colonial and postcolonial periods. These old-growth forests are often located on sites that were too inaccessible or extreme for agriculture or logging, although some larger areas, like the Big Reed Preserve in Maine and Sylvania Wilderness in Michigan, suggest the range of old-growth sites that may have once existed in the region. (Old-growth forests are the focus of Chapter 6.)

THE SPREAD OF PINE AND ASPEN

After the peak of agriculture in the mid-1800s, many farms in the Northeast were abandoned as people moved to better soils in the Midwest and to find employment in cities. Similarly, the waves of widespread harvesting, first beginning in the forests of northern New England and New York and then in the upper Lake States, shifted to the forests of the Pacific Northwest in the late nineteenth and early twentieth centuries. After farms were abandoned and harvested forests left behind, the forest did what the forest is so good at: It regenerated itself.

In abandoned agricultural landscapes, the patches of forest scattered throughout farms across the landscape served as a seed source. It has been estimated that an average farm of the era used approximately 25 cords of firewood per year for heating and cooking purposes. Hardwood species were often harvested from woodlots to fill this need, since conifer pitch and resins are highly flammable and can lead to chimney fires. As a result, a disproportionate amount of conifer species was left in these woodlots and served as a very important seed source in the transition of abandoned pastures back into forests. Most notably, it was a good time to be an eastern white pine.

There are times in history when the circumstances benefit the characteristics and competitive strategies of a certain species, giving that species a chance to greatly expand its population. Acres upon acres of abandoned pasture were left with full sunlight and ample water and nutrients for any tree that could get its roots through the sod. Eastern white pine was the right tree, in the right place, at the right time. Eastern white pine is a wind-dispersed species. Seeds can be carried by the wind more than 700 feet from the tree in open areas, helping the trees spread across the landscape.

The white pine seed also carried with it enough stored energy to fuel the growth of a taproot able to push through the sod and into the soil, providing a reliable source of water and nutrients for the tree's growth. Meanwhile, the seeds of other pioneer species were often too small to carry enough energy to get through the sod and couldn't compete. (Seed from hardwood pioneer species that fell on plowed land fared better, helping hardwood trees remain a component of the regenerated forests.) Once established, white pine was able to take advantage of full sunlight, grow fast, and become dominant. It

Eastern white pine took root in many abandoned pastures in the Northeast in the late nineteenth century.

grew as a dominant species across the region, resulting in a population likely far greater than it would have been without the colonial land use history. Likewise, species such as loblolly, Virginia pine, and tulip poplar regenerated the postagricultural landscapes in the southern part of our region.

The combination of timber harvesting followed by railroad-induced wildfires provided the perfect situation for aspen to become dominant over large portions of the Lake States region. This species, although an important historic component of these forests, was able to take advantage of its ability to reproduce asexually from root sprouts. Many of the railroad fires that occurred following historic harvests happened within a few years to a decade after an area was cut. As a result, the trees growing back in those areas were not sexually mature at the time of the fires, so they were unable to reproduce. In contrast, aspen can resprout at any age, so these disturbances greatly favored its reproduction. Aspen remains dominant in these areas today, with many efforts now focused on trying to bring back the mix of species that was present prior to what is often referred to as the "great cutover."

Creating Homogenized Landscapes

In short, the clearing of forests resulted in a drastic, profound, and uniform change on the landscapes of our region. Clearing acted like a reset button on the clock of succession in a very short period of time. As a result, most of our forests are at the same point in succession, meaning that they have a similar species composition. Many of our forests are also at the same forest development stage and are missing the layers of structural complexity that would normally develop in the wake of small disturbances over time. Many forests now have a lower abundance of conifer species like spruce, hemlock, or pine, either due to their selective removal by historic harvests or their inability to survive the rapid sequence of harvesting followed by fire, particularly in the Lake States. This sameness and lack of diversity makes the forest more vulnerable to today's challenges, as we will discuss in Chapter 8.

Finally, the impact of the genocide and forced removal of Indigenous peoples from across our region cannot be understated. Many woodlands and barrens were historically maintained by cultural burning practices, a feature frequently noted by early European explorers and settlers across our region. The mosaic of forest conditions that colonialism eliminated importantly included these cultural landscapes that supported thriving ecological and human communities, including fire-adapted tree species and animal and plant life dependent on more open savannah and barrens conditions.

WILDLIFE IMPACTS
Beavers, Wolves, and Mountain Lions

The impacts of colonial and postcolonial land use extended, of course, to the many wildlife species that rely on forests. In many areas, the extirpation of beavers had huge and lasting effects. Historically, people have lived in the same places beavers live: low elevation areas near water sources that provided fertile farmlands, drinking water, food, and, eventually, power to operate mills. Beaver wetlands disrupt these benefits. In addition, beaver pelts, prized for their warm and water-repellent fur, became a popular commodity. The combination of these factors resulted in focused efforts to remove beavers from the landscape.

Beavers are known as a *keystone species*. Keystone species have a disproportionate influence on their environment, supporting an entire ecosystem. When beavers create wetlands, they are creating important habitat for many species. Once beavers have exhausted their food source, they abandon their wetland and seek new habitat, and their dam eventually fails, draining the wetland and creating a beaver meadow. The succession clock then begins with the preforest stage of forest development. Grasses and forbs establish, followed by shrubs, and eventually seedlings of trees, providing excellent early successional habitat.

However, these are ephemeral habitats. Succession moves forward. The beaver meadow eventually grows into a sapling-dominated young forest. When the trees are large enough, beavers will move back into

Beaver meadows are critical for species that thrive in open, vegetative habitats.

the area, rebuilding the dam and creating wetland once again. Beaver occupation of a site is dynamic, as they occupy new habitat after using up food sources and as young beavers disperse to establish their own territories. The more beavers in the landscape, the greater amount of these various habitats.

The estimates of precolonial beaver populations combined with firsthand accounts of beavers strongly suggest that these habitats were a widespread, common part of the landscape on which many species depended. Beavers and other disturbances, such as periodic large-scale wind events, resulted in as much as 5 to 10 percent of the landscape being in preforest condition at all times.

The Loss of Apex Predators

While beavers have reestablished themselves in our region at numbers far below their historic populations, other wildlife species have been extirpated from our region with significant impacts on forests. They are our apex predators: wolves and mountain lions. These top predators once roamed the forests of our region, preying on deer, moose, and a variety of small mammals. Wolves and mountain lions held the populations of prey to healthy levels, keeping their browse impact on forest vegetation in check.

Seen as a threat to livestock and humans, these apex predators were aggressively hunted. The combination of hunting and habitat loss eliminated wolves and mountain lions from almost our entire region and significantly reduced their populations in the pockets where they still exist. The loss of these predators coupled with increased deer habitat has meant virtually unchecked deer populations in many parts of our region. Browsing by deer is having a profound impact on forest regeneration and the diversity of understory plants in our forests.

While today's forests are beautiful, they have been heavily influenced by humans. They have become homogenized in age (approximately 60 to 100 years old), species composition, and structure. Our ecologically young forests are missing important components that they simply haven't had the time to redevelop. At the landscape scale, our forests are remarkably similar in many ways as well, and these similarities make them vulnerable to disturbances and pose challenges to wildlife and biodiversity. Ecological forestry seeks to restore missing components to both the forest and the landscape, as well as to restore our relationship to the land.

FIELDWORK

What Is Your Forest's Land Use History?

Our land use history has resulted in homogenized forests of similar age, size, and species. Does your forest have a lot of trees of similar size? Are there areas of your forest with trees of different sizes, either very small or very large? How similar is your forest to the forests surrounding your land?

"When we try to pick out anything by itself, we find it hitched to everything else in the Universe."
—John Muir, *My First Summer in the Sierra*

CHAPTER 3

Forests of Today

OUR REGION HAS SEEN A REMARKABLE REBOUND in forest cover. In fact, it is not an exaggeration to say that the reforestation of our region is one of the greatest ecological recovery stories in the history of the world. The response of our forests is a comment on the resilience of our landscapes—relatively young fertile soils, ample precipitation, and the remarkable regenerative capacity of our tree and plant species drive the recovery of forest ecosystems.

In most parts of our region, forests are 60 to 100 years old. They are old enough to feel like mature forests, providing the essential benefits on which we rely. However, our land use history has left profound and lasting impacts on our forests. Though one hundred years is old by human standards, it is young by forest ecology standards. Our forests continue to make slow and steady progress toward full recovery of all of their complex parts and relationships.

While our forests recover from our land use history, a new suite of challenges have emerged. Understanding the specific ways in which each of these challenges influences forests is imperative to addressing them through the practice of ecological forestry.

A typical second-growth forest of today is homogeneous in species and structure.

CLIMATE CHANGE

Increases in the amount of heat-trapping carbon dioxide (CO_2) and other greenhouse gases in the atmosphere over the last century have led to a phenomenon originally called global warming, now called climate change. *Climate change* refers to the unpredictable nature of current weather patterns and has been appropriately referred to as "global weirding" to capture the unpredictable and unprecedented nature of many of these changes. Climate change has resulted in increased temperatures around the world, including warmer oceans, which alters the way in which our weather patterns operate. Given how important climate is in regulating the growth, survival, and reproduction of the organisms making up our forests, there are a number of anticipated changes to our forests associated with climate change.

Longer Growing Season

Higher temperatures have resulted in growing seasons longer by approximately one to two weeks in our region. This means that forests now have a longer window within which to photosynthesize and make carbohydrates (food), which can mean increases in the forests' productivity, its ability to grow trees and plants. However, all forests have an optimal temperature range in which they are most productive; beyond this range production actually goes down. Higher temperatures also affect the phenology of trees, the timing of the various processes in their life cycle: leaf emergence, seed dispersal, and leaf fall. More specifically, higher spring temperatures may stimulate leaves to emerge earlier, making them vulnerable to a late spring frost.

Tree roots are also vulnerable to increased temperatures. Warmer weather means less snowpack on the ground. Since snowpack insulates the soil and roots, reduced amounts of insulation can lead to root mortality during late frosts and low winter temperatures.

Frequency and Intensity of Disturbances

As discussed in our forest ecology section, disturbances are one of the main drivers of forest succession and structural change. Different disturbances affect different parts of the forest. Some of the most common disturbances in our region include wind events and ice events. In specific parts of our region, fire may also be an important driver of forest change. Higher temperatures mean more energy in the weather system. This increase in energy within the system results in more frequent and more intense disturbances. These disturbances bring a cascading series of impacts.

Wind events range from single tree "tip ups" to hurricanes covering thousands of acres. These types of disturbances usually have the most impact on the main canopy of a forest, those trees with large crowns that catch the wind. More wind events are likely to result in more and more big gaps in the forest canopy. While single-tree tip ups will favor shade-tolerant species, larger gaps in the forest canopy (more than 0.25 acres) will favor shade mid-tolerant species, and

Weather events such as wind and ice storms are predicted to become more frequent and more intense.

gaps of a couple acres in size and larger will favor shade-intolerant species again. These events will increase the amount of snags and downed deadwood in our forests, a critical component that is largely lacking from our landscapes and one of the main differences between old-growth forest and our current second-growth forest. The lack of deadwood in our current forests relative to the historic levels found in old-growth forests will be addressed in greater detail in Chapter 6.

Ice storms, also common in the eastern and central US, cause a buildup of ice on leaves, needles, and branches, resulting in broken branches and even trees snapped off. Higher temperatures may mean that in some parts of the region, precipitation events that may have been snow in the past are now mixes of rain and ice. Given the longer growing season, it is also possible that ice and snowstorms may hit at the end of the growing season while some trees still have leaves. The leaves' surface area will collect more ice and snow than bare branches, leading to more tree breakage and mortality. Like wind events, ice storms often produce snags and downed deadwood in the forest, as well as opening gaps in the forest canopy.

Greater extremes in precipitation patterns are another result of increased disturbances. A deluge can bring an inch or more of rain within a short period of time. These rain events can lead to issues with forest infrastructure such as woods roads, hiking trails, and stream crossings. Care must be given when designing and maintaining such infrastructure to prepare for these large storms and safeguard these investments in the forest.

Between these large rain events, it is common to have periods of drought. Drought conditions predispose trees to decline by reducing the amount of water and nutrients available to them. Like a human who is tired or stressed, trees that are subjected to a drought are weak and have greater vulnerability to disease as well as insects.

The influence of fire on the forests of the eastern US can be described as being on a spectrum. The forests of New England have been called "asbestos" forests. This is because many of the forest communities that are currently present across the New England landscape are not fire-associated ecosystems. This is particularly true in the mesic (moist soil) hardwood and mixed hardwood-conifer forests located in the north, which have very long intervals (centuries) between major fires. In contrast, the pine barren, oak-pine, and oak-hickory forests in New England and New York and across the central Appalachians and Midwest, as well as the jack pine, mixed red-white pine, and boreal conifer forests in the Lake States region, are naturally more influenced by wildfires. It is possible that higher temperatures and drier conditions will favor fire's increased role in the landscape. This is especially true in forest types like northern hardwoods, where

A forest fire created a significant disturbance in this Minnesota forest. Fire is likely to play an increasingly important role in the region as the climate changes.

fire has been extremely rare. Other factors that influence fire include the amount of fuel, especially deadwood, that may be present in the forest due to disturbances such as wind events, ice storms, and insect mortality.

Shifting Composition

Unlike J. R. R. Tolkien's characters the Ents, treelike beings with legs, trees aren't able to walk about the landscape, changing their location. Those that show themselves to be the most well adapted to a site, outcompeting all other trees, will germinate, grow, compete, and live on that site until they die. Since trees are long-lived, we shouldn't expect wholesale shifts in species composition within our lifetimes. However, as trees die in the forest, resources are freed up, providing the opportunity for advance regeneration (saplings under the canopy) to grow and seedlings to establish. Each time this happens there is a competition among individual trees and species to claim those resources and outcompete the surrounding trees.

As temperatures increase and growing seasons extend, species that are at the northern ends of their ranges will expand northward and upslope, away from the warming conditions. Seedlings and advance regeneration are the most vulnerable to lower temperatures and are often the limiting factor for species shifts. Slowly, disturbance after disturbance, canopy opening after canopy opening, the competition for available resources and space will continue, shifting tree species across our region over the next century or two.

More frequent and intense storms should expedite the process as more trees from the main canopy die. Of course, this assumes that species are able to shift their distribution as fast as the climate is shifting. General estimates of past species shifts suggest they move 0.1 to 0.3 miles per year. Unfortunately, climate is currently shifting 4 to 6 miles per year, a rate too fast for most trees to keep up with. Over time, this may generate climate mismatches, where species currently in an area are no longer adapted to the local climate.

Scientists have developed models predicting future habitat suitability for different tree species. However, rather than thinking of these maps as showing where various species will be in the future, it's better to view these maps as predicting geographic areas where species will be best suited and, therefore, the most competitive. Since trees are long-lived, considering future habitat suitability is an important consideration when deciding which tree species to encourage as you practice ecological forestry.

The amount of connectivity within the landscape also plays an important role in tree species migration. Trees rely on wind, insects, and small mammals for pollination and seed dispersal. In a connected landscape, seeds from different species of trees with different establishment and growth strategies can move freely. This diversity of regeneration pathways increases the likelihood that seeds will find sites to which they are well adapted, in turn increasing the likelihood that forests successfully regenerate.

FIELDWORK

How Is Climate Change Having an Impact on Your Woods?

If you have owned your land or lived in the same place for a decade or more, you may be noticing changes related to climate change. Have you experienced more intensive rain events that result in rainfall of more than an inch? Has your forest experienced extended dry periods or even drought? Are you getting the same amount of snowpack and days with frozen soil? Have you noticed your trees leafing out earlier or retaining their leaves longer?

What species do you currently have in your forest? What species may be best adapted to your soils and predicted future habitat?

The fragmentation of landscapes reduces the movement of seeds, making successful regeneration less likely. Therefore, the more connected the landscape, the more resilient the forest.

Finally, those species that are better adapted to drought conditions are likely to compete better in the future. For example, tree species that are able to quickly develop roots that can make contact with mineral soil where water is reliably held will have an advantage. Because shade-intolerant and mid-tolerant tree species have evolved to be most competitive in higher light levels, these species are often well adapted to the drier conditions that come with more sunlight. Practicing ecological forestry includes regenerating species that are well adapted to these future conditions. We will discuss forest resilience more in Chapter 9.

INVASIVE PLANTS

Trees and native understory plants are not the only ones competing to establish, grow, and expand populations within a forest. Over the last several decades, as global trade has expanded, so has relocation of plants, animals, and other organisms from around the world to our region. The forests of our region have faced increasing challenges created by invasive exotic plants, those that threaten to outcompete native species and take over their ecosystems. These plants are native to areas of the world with similar latitudes and that share similar climates and environmental conditions.

Not all plants that are introduced to the United States are invasive. Some are able to persist in the yards and landscapes of our region without being a major disruption to our native ecosystems. However, there are a small percentage that not only survive in their new environment but thrive. Invasive exotic plants are most often introduced to our region for horticulture and landscape purposes; however, agriculture and erosion control have also been sources of these plants. Ironically, landowners were once encouraged and even paid through government programs to plant invasive exotic plants for conservation purposes, including Japanese barberry for a "living fence" and autumn olive for wildlife habitat.

New invasive exotic plants arise periodically as more goods and materials are moved around the world, reshuffling the ecological deck of native ecosystems. Examples of current invasive exotic plants that impact the forests of our region include Japanese barberry, Asiatic bittersweet, glossy buckthorn, burning bush, bush honeysuckle, Japanese knotweed, garlic mustard, mile-a-minute weed, Japanese stiltgrass, and invasive honeysuckles.

Not having evolved over thousands of years with other plants, animals, and insects of our region, invasive exotic plants don't have any natural predators. They are fast-growing, allowing them to use the increasing temperatures and carbon dioxide of the atmosphere to rapidly spread. Like many understory plants, invasive exotic plants leaf out early in the spring, before trees do, and intercept the sun. A longer growing season, especially in the spring, provides these plants increased opportunities to grow. Many of these plants are also prolific seed producers, allowing them to spread easily. Their strategies are very well suited to taking advantage of disturbed sites, those with a lot of light, increased temperatures, and exposed soils. Once established, the shade they cast and, in some cases, the allelopathic (harmful) chemicals exuded through their roots inhibit the regeneration of other species. Together these strategies make these species highly competitive in our region.

Like many invasive plants, Japanese barberry leafs out early to take advantage of sunlight coming through the canopy. It can dominate the understory, preventing the regeneration of native understory plants and trees.

A Simplified Forest

The most significant problem with invasive species is their ability to outcompete regenerating species of both trees and understory plants, leading to understories dominated by invasive plants that are very difficult and expensive to remove. Forest understories dominated by invasive exotic plants simplify our forests, increasing the homogenization that resulted from our land use history—the opposite goal to that of ecological forestry, which focuses on increasing forest complexity. The trees in our current overstory are likely to live decades or even centuries, so the impacts of these regeneration challenges may not be seen for some time. However, when those trees in the main canopy do begin to die, or when one of those natural disturbances that are predicted to increase in frequency and intensity impacts a forest, a lack of regeneration will have profound impacts on the forest and the benefits it provides. Ensuring tree regeneration has become a critical issue across the region.

The establishment of invasive exotic plants in our forests creates *novel ecosystems*, combinations of trees and plants that have never grown together. The ecological and economic impact of these novel ecosystems is significant. Since our wildlife species haven't evolved with invasive plants, fewer of our native wildlife species use the plants. For example, fewer native insects can be found on exotic invasive plants, and deer prefer to eat native plants. In addition, invasive exotic species tend to have fruits that have less nutritional value to wildlife. In fact, the fruits of invasive exotic species have been likened to "junk food" for wildlife. From a human health perspective, species like Japanese barberry provide cover for rodents that carry ticks, increasing the risk of disease transmission to humans.

Human-caused disturbances, lack of complexity, and an absence of native species that fill the same ecological niche as the invasive exotic plants are some of the factors that make a forest more vulnerable to invasion. Once established, invasive exotic species are spread in a number of ways. In terms of sexual reproduction, the seed of invasive exotic plants can be spread by animals, such as birds who eat their fruit and then drop their seed, and wind. From an asexual regeneration perspective, invasive exotic species often have an aggressive root system that spreads rapidly and can outcompete other roots in its vicinity, as well as an ability to resprout prolifically when cut. Humans can also spread seed on our hiking boots and clothes, on machines that travel through the woods (such as ATVs and timber-harvesting equipment), in the fill that may be brought onto a site, and within hay bales that may be used to stabilize an area.

Controlling Invasive Plants

The control of invasive exotic plants is time-consuming and expensive. The most effective strategies are likely tighter safeguards on the movement of plant species through trade. At the woodlot scale, the most cost-effective approach is to monitor your forest for invasive exotic species and remove them before they become well established and difficult to treat. Hand-pulling a few individuals is far better than dealing with an acre or more of well-established plants. Control of established infestations usually requires a chemical treatment with herbicides, which can understandably be a concern to landowners and can bring undesirable environmental and human health risks, depending on the chemical and how it is applied. However, at this point it is the only effective strategy. In the absence of treatment, the invasive exotic plant will likely continue to grow and spread, limiting a forest's ability to function properly.

Working with a forester to address the infestation can be a helpful approach to invasive exotic plant control and is essential before implementing any timber harvesting.

Now, to be fair, our forests are impacted not only by invasive exotic species. Some of our regeneration challenges are the result of invasive native plants taking over forest understories and making regeneration very challenging. Increases in invasive native plants are usually due to changes in land use, disturbance regimes (e.g., fire suppression), or levels of deer browsing that preferentially favor the growth of these species, generating profound impacts on our forests. For example, hay-scented ferns can create dense shade, a thick organic layer, and cover for rodents, all of which can limit tree seedlings in a forest. Mountain laurel, while a beautiful flowing shrub, can likewise dominate an understory, making regeneration of tree species challenging. Grapevines work their way up trees and into the overstory. The vines take over crowns, limiting a tree's ability to photosynthesize, making it dangerous to fell, and, in some cases, weighting it down to the point where it will break. These native invasive plants are also important to account for in forest stewardship.

Looking at a forest, it is common for people to see green plants and assume everything is okay. Like the evils let out of Pandora's box, once they are unleashed, invasive exotic plants cannot be completely eradicated from our landscapes. Although the control of these plants is an expensive and time-consuming endeavor, their detrimental impact on forest regeneration, and therefore long-term forest health, complexity, and resilience, makes their control essential in both active and passive forest management approaches. Invasive species control helps a forest regain its function and fully express itself. More information about their control will be discussed in Chapter 9.

FIELDWORK

Are Invasive Exotics Present in Your Forest?

Find out what invasive exotic species are common in your area by talking to your local service forester or extension forester (see Resources on page 264 for contact information). Learn to identify invasive species and look for them in your forest. Monitor your forest to catch emerging small issues. Remember that the sooner they are identified, the better chance you have at eradicating them. Work with neighbors to control invasive exotics in your landscape and reduce the risk of spread back into your forest.

INVASIVE INSECTS AND PATHOGENS

Plants are not the only nonnative threat to our forests. The same global trade that brings invasive exotic plant species to our region also moves nonindigenous, invasive pests, often referred to as invasive insects. Invasive insects typically enter the United States in wood material used for shipping or on live plants. Like invasive exotic plants, invasive insects from other continents such as Asia can spread here due to the similar climate and vegetation, given our common latitudes. Once here, these invasive insects can escape into the natural environment.

For thousands of years, trees and forest insects have evolved together. Forest insects often feed on tree leaves, needles, and cambium (inner bark). Trees provide habitat for insects within the nooks and cracks of their bark, in the crown, on branches, and even in the litter of the forest floor. In response to forest insects, trees have evolved physical and chemical defenses. Examples of physical defenses include the texture of leaves, small appendages on the surface of leaves (called trichomes), thorns, and resin ducts. The sugar maple has evolved a tough sheath over the veins of its leaves to protect its carbohydrates from insects, and small glasslike crystals in its leaves to dissuade caterpillars. These defenses are often concentrated in trees when they are young and most susceptible to damage. Plants have also evolved chemical defenses, including repellents, toxins, and chemicals that reduce digestibility, or they discourage insect attacks with pitch or resins, and some have even evolved to lack the very nutrients that insects seek out.

These characteristics help the tree survive and pass on its genetic code to future generations. The characteristic then becomes common in the population, lasting until the insects adopt a different strategy, which the tree then has to deal with. If an individual in the population has a characteristic that makes it more fit, it survives and passes it on. Today's trees are the result of generations on generations of evolution that have produced the trees most well adapted to face the environmental conditions and challenges in a given habitat.

However, because invasive insects haven't evolved in our region, this long evolutionary process has not had time to play out with them. If the tree species on which an invasive insect feeds doesn't have a resistant characteristic, and there are no environmental limiting factors, such as low temperatures, the insect will move through the population of defenseless trees, killing many of them.

The Ripple Effect of Species Loss

Some current invasive insects impacting the forests of our region include the hemlock woolly adelgid, native to East Asia, that feeds on the sap of hemlock. The emerald ash borer, from northeastern Asia, feeds on the cambium of ash trees. The Asian longhorned beetle, native to Korea and China, feeds on the living tissue on the underside of the bark, with a preference for a number of hardwood species including maples, birch,

Hemlock woolly adelgid

Asian longhorned beetle

Emerald ash borer

horse chestnut, willow, and elm. The spongy moth (formerly gypsy moth) feeds on oak leaves. There are also numerous pathogens, including the ones that cause beech bark disease, oak wilt, butternut canker, and chestnut blight, that are having a significant impact on our forests. Since invasive insects and pathogens are being introduced into the US regularly, it is only a matter of time before a new one emerges in the natural environment.

As discussed in Chapter 1, different disturbances impact different parts of the forest. In the case of invasive insects and pathogens, their damage is highly selective, often targeting trees of a certain species. We are very fortunate in many parts of the region to have a diversity of trees within our forests, so the land often remains forested after the invasive insect or pathogen has killed its host trees. However, once again, the forest structure and species composition have been altered and simplified. Other species that use the tree are also threatened, particularly those that have evolved to feed exclusively on that species.

Beyond these ecological impacts, the loss of a species from our forest also has cultural and ethical implications. Though tree mortality is a good and normal process in a forest, excessive tree mortality is reason for concern, particularly as we increasingly look to our forests to mitigate climate change and meet societal demands for wood products. The carbon gains made by forest growth and development can be outweighed by the mortality caused by invasive insects, a topic covered in Chapters 7 and 9. Finally, the mortality caused by invasive insects can influence the very functions of the forest itself, such as nutrient cycles and the hydrology of a forest. The loss of hemlock, a species that often grows along streams, means these trees will no longer shade the habitats of cold-water fisheries and no longer drop organic matter such as needles and wood into the water, thus threatening the foundation of the aquatic food web.

Culturally, Indigenous peoples have deep relationships with particular species. These species may be a part of their creation story, a traditional food source, or tied to a cultural practice. Black/brown ash is an excellent example. Found primarily in wetlands and along streams, black/brown ash is the primary source of material used by Indigenous basket makers. Ash is also a part of a number of Indigenous creation stories. Losing ash trees causes those cultural ties to unravel. Preventing a species from going extinct is also an ethical obligation, a responsibility to maintain the full range of native biodiversity and to honor the gifts that a species provides us.

The northeastern and midwestern US have the highest concentrations of invasive insects in the country. This is due to the amount of entry points from global trading, such as ports and major airports, and to the number of ready hosts. Once introduced, the dispersal rate of an invasive insect is dependent in part on its ability to move through the landscape and the concentration of target species in the landscape. Dispersal rate is also dependent on human actions. Trucking logs to be milled into wood products or firewood is a common way in which invasive insects are dispersed across the region. To reduce spread, states may impose quarantines on the movement of certain tree species for the parts of the year when the insect is dispersing.

The infestation of invasive insects can feel overwhelming. However, there are adaptation strategies that can be applied through ecological forestry to help forests move through these threats and even to preserve threatened species on the landscape. These strategies will be discussed in Chapter 9.

EXCESSIVE DEER BROWSE

One of the top goals for many landowners is supporting wildlife. Therefore, it may seem contradictory that one of the threats that our forests face is from a species of native wildlife: white-tailed deer. Deer have a long evolutionary history with trees and forest plants. When deer populations are at healthy levels (about 5 to 15 deer per square mile), the relationship between deer and the forest is in balance. However, if the deer population increases, deer overpopulation can have a negative effect on forests. Over the last several decades, deer populations have increased in many places within our region, resulting in chronic overbrowsing of the forest and leading to a failure of forest regeneration.

There are two primary reasons for increases in deer populations. First, deer eat a wide variety of vegetation and range over large areas, meaning that deer can subsist and even thrive in a broad range of habitats. For thousands of years, the hunting of deer by Indigenous peoples and by apex predators, like wolves and mountain lions, helped keep deer populations in check. With European settlement, there was increased pressure on deer for both their meat and hides, leading to significant reductions in the population, which then rebounded as Indigenous populations in the region declined.

The apex predators that historically helped keep deer populations in check have, with very few exceptions, been extirpated from our region. Deer hunting, which is viewed by some as a substitute for the control of deer densities provided by apex predators, has also declined over the past few decades. Finally, in some areas, the spread of suburbia, complete with well-landscaped yards of tasty shrubbery like cedar, has increased deer habitat and food sources. Deer have responded across our region by increasing their populations.

In the absence of apex predators, deer hunting has been the primary control on populations. Though deer hunting remains a common activity in parts of our region, the overall number of deer hunters has steadily declined over the last few decades. This has allowed deer herds to swell. The impact of deer overpopulation on our forests has been significant. Estimating the amount of deer pressure on your forest is an important consideration in how to steward your land.

How to Estimate Deer Pressure

There are several options for estimating deer pressure on your forest. Some methods are quantitative and can estimate the actual number of deer. Other methods are qualitative and ask the observer to provide a rating of low, medium, or high deer pressure based on observation. Your goal is to get a sense of how many deer are in your landscape by looking for the physical signs of their presence.

Though deer have preferences among the plants they browse, as herbivory generalists they have a broad palate. Therefore, the first plants in an area to be browsed are often preferred species, including:

- **Woody species:** Northern white cedar, hemlock, sugar maple, oak, hickory, yellow poplar, and black gum
- **Plants:** Pink and yellow lady slipper, trout lily, white wood aster, New England blazing star, Canada lily, Canada mayflower, trillium, and ginseng

Once the preferred plants have been browsed, deer will move to other less palatable plants until finally, with no other options, they will browse on the least preferable species, including:

- **Woody species:** Balsam fir, striped maple, black birch, American beech, jack pine, red pine, white pine, black locust, tamarack, and American holly
- **Plants:** Garlic mustard, jack-in-the-pulpit, common milkweed, Pennsylvania sedge, blue cohosh, swallowwort, mayapple, mile-a-minute, stiltgrass, buttercups, and little bluestem

If deer are browsing on these low-preference species, it typically means that the swollen population has already eaten all the high-preference species.

Another way to estimate deer populations is by counting scat piles in your forest. There are protocols that you can find online to do this, but it typically involves walking transects through the forest, counting the number of deer scat piles, and then using an equation to turn the number of scat piles into a population estimate.

Finally, you can train your eye to observe the impacts of deer. A healthy forest should have multiple layers of vegetation with both tree regeneration and herbaceous plants in the understory. A four-foot-high *browse line* with nothing growing below it is a strong indicator that there is heavy deer pressure. Heavy deer pressure can also lead to understories dominated by invasive plants and ferns, which deer don't browse. Tree regeneration and stump sprouts that show signs of browse (the end of the branch is broken off and there is no bud) are also indicators. Looking for these signs of deer browse can help you determine the extent of deer pressure.

Loss of Biomass

As discussed in Chapter 1, as forests move through succession they increase in complexity, gaining trees of different sizes and a diversity of vegetative layers. This leads to an increase in biomass—the amount of plant material in the forest. The impacts of excessive herbivory (the consumption of plant parts by animals) can be described as *retrogressive* succession. That is, instead of the forest getting more structurally complex over time, it becomes more simplified. Likewise, instead of the forest gaining biomass as it moves through successional stages, the forest actually begins to lose biomass. The loss of biomass from the forest means less carbon stored in it, reducing

An understory heavily browsed by deer severely limits the forest's ability to regenerate itself and grow more structurally complex through time.

the forest's ability to help mitigate climate change.

It is important to remember that the home range of a deer is about 1 square mile (650 acres), so unless you own a very large piece of land, the deer moving through your forest are also moving through the forests of your neighbors. If your land is experiencing intensive deer browsing, it's likely that your neighbors' is, too. Addressing the problem at a landscape level by working with your neighbors to control the local deer herd or sharing effective strategies for protecting regeneration will likely be the most effective approach.

It's not only deer. In the northern part of our region, moose populations can also be quite high. An adult moose can browse around 40 to 50 pounds per day, having a significant impact on forest regeneration. Moose are also taller than deer, so a young tree that is about 4 to 5 feet tall is typically safe from deer browse, but a tree must grow until it is 8 feet to be safe from moose browse.

FIELDWORK

Is Your Forest Heavily Browsed by Deer?

Look for signs of excessive herbivory. Does your forest have a browse line? Does your forest have tree regeneration established? Is your understory dominated by ferns or invasive plants? Is the tree regeneration browsed? Are you seeing a lot of deer scat piles?

FOREST PARCELLATION AND CONVERSION

While the previously discussed challenges are all ecologically based, perhaps the most pressing issue our forests face is that of parcellation and conversion. The mostly forested landscapes of our region are like a quilt of ownerships, most of them held by families. The decisions that families make about the future ownership and use of their land will lead to either the land being passed on as a forest or the land being divided and converted to another use, such as house lots.

Dividing Up the Landscape

Since colonial times, landownership in the United States has represented freedom, power, and financial security—a tradition that led to the taking of land from Indigenous peoples, which continues today. Land is a tremendous, flexible asset that provides many social and economic advantages. When transferred from one generation to another, land accumulates wealth and provides a head start for the next generation. It can help landowners access financial and social opportunities that non-landowners may not have. Therefore, it is not surprising that many parents want to provide their children with these advantages.

The current size of a property is the result of the decisions preceding landowners have made over the last two hundred years. A common tradition was to give land to the oldest son to help keep the property intact as a viable farm. As agriculture declined in importance as the main source of livelihood for many families, and as social norms regarding gender evolved, parents began giving land to all their children, male and female. Since land is highly valuable for its residential and commercial development potential, it is also common for landowners (or their children once it's inherited) to sell off portions of their land to generate income. Either way, the result of these decisions is often smaller and smaller parcels of land. This process of dividing land is known as *parcellation*.

Parcellation complicates our landscapes and impairs our ability to ensure forest benefits. As discussed in Chapter 1, the many public and personal forest benefits that we seek to conserve exist at scales far beyond the individual property level. The more people owning the land, the more decision-makers must be reached with information, and the more people needed to conserve the resource(s), each of whom has their own unique set of goals and financial and social circumstances.

For example, wildlife habitat for an endangered species may require some form of active management. Reaching 50 landowners who own property within the habitat of that endangered species, teaching them about that species, and providing them with stewardship options to meet their own goals while protecting the species is far harder than reaching 10 landowners. A hiking trail is another excellent example. If a regional trail goes through the properties of 10 landowners, all of whom give permission for the trail to cross their land, and one landowner divides their property, selling it to five additional landowners, the snowmobile/biking/hiking organization that stewards that trail will now need to reach agreements with each of those landowners. If even one landowner refuses the trail agreement, it can provide a significant challenge to the continuity of the trail and its benefits.

Development and Conversion

While parcellation threatens the innumerable benefits our forests provide, the conversion of forests to non-forest land uses (such as residential development) eliminates forest benefits altogether. Forest conversion can happen in a number of ways, the most common being conversion to low-density residential development. People need places to live, and over the last several decades there has been a general trend of people moving from urban to suburban areas. In fact, for many, there is a perception that being an environmentalist means getting back to the land and being independent. Ironically, well-planned cities, where resources can be concentrated, make more environmental sense than suburban sprawl, which eats up both forest and farmland at a rate disproportionate to the actual rate of population growth. While some forests in our region have very low development pressure, other areas, particularly those within commuting distance to population centers, and areas with employment opportunities, have high pressure. These "tension zones" around urban areas result in increased parcellation and forest conversion. No state is immune to the effects of forest conversion.

Converting forests to other land uses eliminates the many benefits they provide.

While forest conversion due to suburban sprawl is a decades-old trend, there are recent political and social influences that have sped up this trend. After the September 11 terrorist attacks, our entire nation was shaken, and for good reason. The attack on US soil made us feel vulnerable, and people began to reconsider where they lived, resulting in an increase in people moving out to rural areas. Likewise, the COVID-19 pandemic turned our world upside down. The assumptions and norms we had all been living under came into question. It became possible in many professions to work remotely. Similarly, this accelerated a movement to rural areas.

As our need for energy increases, so do efforts to reduce dependence on oil, which are putting increased pressure on the forests. Solar fields, for example, have become an additional cause of forest conversion. While few would argue about the importance of renewable energy, it has resulted in the conversion of many acres of forests and the loss of the many benefits they provide.

As discussed in Chapter 2, this is not the first time that we've lost forest cover. In the nineteenth century, many of our forests were cleared for agriculture. The good news is that, in many parts of our region, agricultural land returned to forest. Because this change from forest to agriculture was significant but not permanent, people refer to this deforestation as a *soft change*.

However, the conversion of forests to residential and commercial development is referred to as a *hard change*, since these changes are more permanent and involve activities like paving, building structures, and altering landforms through excavation. This is not to say that areas that are now developed cannot go back to being natural areas. In fact, there are examples of people razing structures to restore natural systems. However, the reality is that when forests are converted to residential and commercial uses, it represents at least the long-term, and maybe permanent, loss of forest benefits. Further, since the restoration of forests after soft changes is taking hundreds of years, restoration after hard changes will likely take even longer.

SYNERGY AMONG CHALLENGES

It's easiest to explain each of these challenges individually, as if they occurred in isolation. However, everything in our forests is connected. To gain a full understanding of these challenges, it's important to understand their relationships and synergy.

Invasive plants are helped by several other challenges. Climate change provides many advantages to invasive plants, which are disturbance-adapted species that take advantage of new openings in the forest canopy. Increases in the frequency and intensity of disturbances, such as wind and ice storms and outbreaks of invasive insects, will likely result in more of these openings. Invasive exotic plants also benefit from extended growing seasons and higher temperatures in the early spring before the forest overstory has leafed out. Increased carbon dioxide in the atmosphere acts like a fertilizer, giving fast-growing invasive exotic plants more raw material to make energy through photosynthesis.

Deer prefer to browse the native plants with which they have evolved, giving nonnative plants another competitive advantage. In areas of high deer browse, the forest structure is simplified, providing opportunities for invasive plants to establish in the niches vacated by native plants that have been overbrowsed. At the same time, milder winters result in lower rates of deer mortality, increasing deer herds and browse pressure in these forests. Recent research is also suggesting that areas with high deer density may also host higher amounts of invasive earthworms that can lead to understory plant layers dominated by one or a few native or nonnative plant species.

Finally, the parcellation and conversion of forests creates more edge and less interior forest. Invasive plants typically do well in these edge areas of high disturbance and light. In addition, homes themselves can be sources of invasive plants, due to the landscaping industry.

Invasive insects have also spread thanks to climate change. Low temperatures can be a significant limiting factor to the spread of insects. For example, temperatures of –10°F to –20°F (–23°C to –29°C) will kill a significant number of hemlock woolly adelgid. It was believed that when this insect first became established in the northeastern United States that cold temperatures would limit its spread. As the woolly adelgid moved north through Connecticut, Massachusetts landowners hoped that their typically slightly colder winters would stop its advance. However, warmer winters resulted in its continued spread. Likewise, there was a time when landowners in Maine considered the woolly adelgid a forest health issue of southern New England. Now that hemlock woolly adelgid is present in Maine, there can be no denying that climate change has aided this invasive insect.

Human Population Changes

Climate change has had an influence on forest parcellation and conversion as well. A growing trend is the phenomenon of "climate refugees." Increasingly, climate change is going to make certain areas within the US and across the world difficult or even impossible to live in. As sea levels rise, millions of people may be displaced from areas along the coast (such as Florida). Increased drought and frequent air quality issues tied to wildfires may make areas of the southwest and western US difficult to live in. Meanwhile, the Northeast and Midwest are predicted to be desirable regions, given our temperate climate and abundance of fresh water. Increases in the population of our region will result in greater pressure on our natural resources.

The edge conditions created by forest parcellation and conversion provide favorable habitat conditions for deer, who are a generalist species. In addition, increased housing density results in fewer opportunities to hunt deer, since hunting safety regulations typically restrict how far one must be from a residence in order to hunt. Increased parcellation results in more landowners from whom one must gain permission to hunt, and increases the chances that the land will be posted as private,

Our forests are resilient, but restoration takes hundreds of years.

prohibiting hunting. As rural areas change to more exurban and suburban areas, there is often a shift in attitude away from hunting, resulting in fewer hunters. At the same time, residential development often brings with it landscape plantings that provide a food source for deer.

As John Muir once wrote, "When we try to pick out anything by itself, we find it hitched to everything else in the Universe." All things are connected. These challenges to our forests cannot be considered in isolation. To be successful, we must recognize their connections and implement strategies that address them together. Ecological forestry provides us an opportunity to address these challenges by considering the whole forest and focusing on restoring its missing pieces, encouraging it to fully function and helping it find ways to adapt and move through these many challenges. As a landowner, you play an essential role.

"Between the two extremes of passively following nature on the one hand, and open revolt against her on the other, is a wide area for applying the basic philosophy of working in harmony with natural tendencies."

—H. J. Lutz, *Forest Ecology, the Biological Basis of Silviculture*

CHAPTER 4

Tending the Future with Ecological Forestry

OUR FORESTS HAVE SHOWN THEMSELVES to be remarkably resilient. They have rebounded from many natural disturbances, such as hurricanes, tornadoes, and fires; agricultural clearing; intensive logging; and invasive diseases, like chestnut blight, while being shaped by them along the way.

Will today's forests be able to keep pace with the increasing number and frequency of challenges? Ultimately, their inherent resilience will likely carry them through, but how long will it take for them to adapt? What type of species and structure will they have? And, most importantly from a human perspective, will they continue to provide the types of benefits on which we rely? After all, humans need forests. Forests do not need humans.

Ecological forestry applies the principles of forest ecology to sustain people and forests.

Our forests have made a remarkable rebound across our region. Now it is time to put into place the strategies to steward them effectively and the relationships to sustain them long into the future.

A WAY FORWARD

In forests where current challenges are relatively less intense, we can have confidence that growth and change are substantially driven by historically natural processes and will move toward the endpoint of an old forest, with all of its wonderful complexity.

However, in forests facing a high intensity of challenges, there is good reason, backed by abundant field-based experience, to suggest that the strategies that forests have developed over thousands of years to sustain themselves have been less successful. In fact, for the first time, there is growing concern among the forestry community in our region about our ability to successfully regenerate our forests. That's not to say that forests will not regenerate, but they may not fully regenerate or lead to the types of forest species and structures (and therefore function) that we have had in the past. In other words, the trajectory of forests through time that we spoke about in Chapter 1 is not the same trajectory that forests facing current challenges are likely on. If left alone, they may not reach an old forest condition as we currently define it. We have entered a new era.

Looking back over our colonial land use history to today, it would be easy to say that the problem with forests is humans. Humans are responsible for the wholesale clearing of forests for agriculture, industry, and development. Humans removed wood products from the forest in an extractive manner with little effort directed toward forest restoration. Humans introduced invasive insects and plants. Humans have caused climate change.

There is no question humans have had a profound, often negative, impact on forests. It is, therefore, easy to understand the rationale of those who advocate for the separation of humans from forests through policies that seek to eliminate active forest management.

While the reaction is understandable, it is impossible to entirely extricate ourselves from the forests due to our reliance on the essential benefits forests provide us. Separating ourselves from forests fails to address the core issue, which is our relationship with forests. While wildlands are an essential component of ecological forestry, we need to find a balanced, ethical, reciprocal relationship with forests. Without this new balance, we will never reach our goals.

The future of our forests can feel uncertain, even depressing. It is not an exaggeration to say that our very survival depends on forests to provide the clean water we drink, the air we breathe, the wood products we use, and the climate change mitigation on which we depend. Forests help us meet our ethical responsibility to ensure the full range of biodiversity is conserved, as well as other important life-sustaining benefits such as recreation and spiritual renewal. If we do not address these challenges, we are essentially walking away from our responsibility to steward the forests providing so much benefit to us.

Stewarding forests and ensuring their health and ability to function properly is our responsibility. Practicing ecological forestry provides opportunities to do just that—it is a way forward. There are ways through the practice of ecological forestry to not only address these challenges but also meet your own goals for the land, continue to produce public good, and, in the process, change our relationship to forests. You play a profoundly important role in applying these strategies. You have the opportunity to apply these principles in a unique way that matches your goals and forest and to share what you learn with your friends and neighbors.

AN EVOLVING RELATIONSHIP WITH FORESTS

Before outlining the basic principles of ecological forestry, it's important for us to understand that forestry in the US has evolved over time. First, we must acknowledge that forest management in our region didn't start in the colonial era. Indigenous peoples were managing forests long before the arrival of colonists. The exploitive harvesting practices applied in the nineteenth and early twentieth centuries led to concerns that our once seemingly inexhaustible timber supply was not being managed to ensure a future supply, which could lead to a timber famine. In response, forestry started as a science-based approach adopted from Europe and focused on the sustained yield of timber to fuel the wood needs of a growing country.

As affluence increased in the twentieth century and people started engaging in leisure activities, our perspective on forests changed. Forests became more than the source of our wood products—they were places to find respite. They began to be valued for a broader diversity of benefits, such as wildlife, water, and recreation, leading to the concept of multiple-use forestry. The 1960s and 1970s brought the environmental movement, inspired by Rachel Carson's seminal book *Silent Spring* and events such as the first Earth Day. In the 1980s, the concept of ecosystem forest management, which placed value on all the parts of the forest, began to take hold. This evolution of forestry reflects changes in our understanding of forests, the shifting attitudes of society toward forests, the rise of new critical issues, and the consideration of diverse worldviews and values. The evolution continues today.

Though forestry's reputation carries the scars of its past, there is much to be proud of when it comes to our current forestry practices. Standard forest management, sometimes called "timber-focused management," is the norm across our region and is responsible for the sustainable stewardship of millions of acres of forests, providing tremendous public benefit. However, as our understanding of forests and their challenges evolves, so does our need to continually refine our strategies to steward them.

Although standard forest management does take into consideration the influence of forest management activities on certain aspects of forests, like soils and water, it continues to have a strong emphasis on

SILVICULTURE

Achieving Your Desired Forest Conditions

Foresters use the term *silviculture* to describe standard forest management strategies to create the forest conditions that will help you reach your goals.

Silviculture is the science-based art of establishing, growing, and regenerating forests to meet landowner goals and the needs of society.

Silvicultural systems are a planned series of forest management strategies that lead to a forest with the desired conditions—structure and species composition—to meet landowner goals and societal needs while sustaining forest health and function. Silvicultural systems are named for the way in which the forest is regenerated. For example, shelterwoods, seed trees, and clear-cuts are intended to establish even-aged forests comprised of trees that are mostly the same age and typically shade mid-tolerant or intolerant, while selection systems aim to establish and maintain forests with three or more age classes of trees that are shade mid-tolerant or tolerant.

management activities that generate the greatest wood outputs. The application of ecological forestry shouldn't be seen as a wholesale abandonment of standard forest management. Instead, ecological forestry is meant to use the proven strategies of standard management while focusing on maintaining all aspects of your forests, including those that produce wood, like crop trees, and those that serve critical ecological functions, like downed deadwood, large, old trees, and the diversity of living things that can't be harvested and put on a log truck.

Ecological forestry is centered on the respectful and reciprocal engagement with forests in ways that support the diversity of benefits they provide. As such, it honors the traditional stewardship models established by Indigenous communities throughout the region covered by this book and integrates those with the many contemporary challenges and demands facing our forests. Although elements of ecological forestry can be found in scientific articles dating back over a century, and within tribal communities for millennia, the formalization of the principles and practices tied to this approach of forest stewardship has been quite recent and is largely due to the influence of Jerry Franklin and Norm Johnson's book *Ecological Forest Management* and Brian Palik and his coauthors' book *Ecological Silviculture*. We've adapted much of their work to provide principles and practices to guide respectful engagement with your forest.

We must ensure that forests are whole and properly functioning ecosystems. By focusing our efforts on restoring their missing components, we can make sure that forests are equipped to meet the challenges of today and will continue to produce the essential benefits on which we rely. In doing so, forests can be restored to full versions of themselves. This doesn't mean they will be identical to the forests of the past but rather that they will be able to express the full complexity and diversity inherent within them and have the capacity to continue to change and evolve. This is the focus of ecological forestry, and it is the next evolutionary step in forestry.

PRINCIPLES OF ECOLOGICAL FORESTRY

The overarching emphasis of ecological forestry is to model forest management on natural forests themselves in order to sustain and restore all parts of the forest. As covered in Chapter 1, forests have evolved strategies over thousands of years to establish, grow, and develop within the pattern of disturbances of our region. Ecological forestry uses these natural processes as models, including for the emulation of natural disturbance. This allows forests to express the strategies they have developed through time in response to these types of events. Using natural disturbance to define our management practices recognizes that forests have boundaries and limits that can help determine the amount of trees we harvest, the frequency of harvests, and, just as important, what we leave behind.

Over the past several decades, four general principles have been developed for guiding the application of ecological forestry. These principles help identify strategies within standard forest management that can be emphasized, others that can be de-emphasized, and some new strategies that can be adopted.

Principle 1: Continuity

One way that ecological forestry strives to model natural forests is by purposefully retaining elements from the current forest into the next stage of the forest. More specifically, the principle of continuity emphasizes the deliberate creation and maintenance of biological legacies, particularly after timber harvesting. Biological legacies are the threads from former iterations of the forest that are woven into new versions of the forest, providing powerful mechanisms for resilience and recovery. Deliberately leaving biological legacies, such as older trees, snags, and downed trees, helps to regenerate forests after a timber harvest and sustain other benefits such as wildlife habitat and carbon storage.

A commitment to continuity is a commitment to not harvesting trees with the same intensity and frequency as with standard forest management. The idea of limiting timber harvesting to some degree, as determined by landowner goals, can seem contradictory to forest management, which is often assumed to be fully centered on eventually harvesting trees. However, by using nature as a pattern, ecological forestry recognizes the important relationship between areas of the forest that have been disturbed and other parts of the landscape that have not. These passive and active approaches are not in conflict, but in concert.

The principle of continuity must extend beyond your forest and into your home, as the decisions you make about future ownership and use of your land will have the most profound impact on your forests. There can be no continuity within the forest if the forest is converted to another land use, such as a housing development. Ensuring that the land stays in its forest condition provides the time necessary to develop the missing components and continue providing benefits to the future landowner and society. Conservation-based estate planning provides an opportunity to ensure the continuity of forest cover and management goals through a formal plan (discussed in Chapter 12).

Principle 2: Complexity and Diversity

Forest ecosystems are naturally complex and diverse, producing a vast array of benefits. Many of the benefits are ones that we know and value. There are also likely some benefits we do not even yet recognize as important. Managing forests to optimize a single outcome, such as carbon sequestration or timber, results in the reduction of diversity within the forest and the simplification of forest structure, leading to fewer overall benefits and destabilizing the forest ecosystem.

STANDARD FORESTRY

Standard forestry dedicates almost all of the area within a forest to active forest management over time. A small amount of the forest (less than 20 percent) is designated for retention of forest structure and species diversity. The illustration above represents approximately 5 to 10 percent of the area dedicated to those purposes.

ECOLOGICAL FORESTRY

Ecological forestry dedicates a portion of the forest to active forest management but designates a substantial portion (20 to 50 percent) of the forest for the retention of forest structure and species diversity to achieve forest continuity and complexity over time. The illustration above represents approximately 40 to 50 percent of the area dedicated to those principles.

KEY

 All current dead wood is retained

 Noncommercial species retained to increase biological diversity and forest resilience

 Legacy trees retained to develop into large, old trees and future deadwood

 Trees available for removal from the forest

The history of forestry demonstrates this principle very well. In the early years of forestry, producing wood products was the sole focus. Treating forests the same as an agricultural crop like corn resulted in diminished benefits and less resilient forests. Today, we have far more people depending on our forests for their many benefits. Forests also face far more threats. Strategies that lead to less diversity and simplified forest structure reduce the ability of the forest to produce and sustain the full suite of forest benefits into the future and make them vulnerable to the threats they face.

Ecological forestry explicitly recognizes and celebrates the promotion of complexity and diversity at both the forest and landscape level. This is achieved by encouraging a diversity of tree species by creating conditions that promote forests at different points on the successional clock, as well as by supporting all stages of forest development across the landscape. Heterogeneity builds resilience by ensuring that all forests and parts of forests are not susceptible to the same disturbances at the same time. Having forests at various stages also complements the way in which our wildlife species have evolved, with the majority of species depending on multiple age classes in the landscape.

Principle 3: Timing

It is only natural for landowners to make decisions about forest management based on human time. The events of one's life, such as retirement or the need for a new roof, can be a trigger for a timber harvest. Another example of a decision based on human time would be planning a timber harvest based on a tree's financial maturity, the point at which it achieves potential for a specific financial return (discussed in Chapter 11).

Rather than focusing forest management decisions on human time, ecological forestry encourages forest decisions based on forest time. These longer time frames allow for the development of important forest characteristics that would likely never have the opportunity to develop on human time. For example, if forests were always harvested at their financial maturity, trees wouldn't have the chance to grow as they do in natural forests. Trees wouldn't have the time to develop complex bark and cavities that are so important to wildlife and biodiversity. Forests would never have the opportunity to accrue their maximum amount of carbon. Understory plant communities, which typically spread very slowly, wouldn't have time to reestablish themselves.

And it isn't always about waiting on time. Sometimes making decisions on forest time is about speeding up time by accelerating the development of complex structure within the forest. Of course, adhering to this principle can require a trade-off between the ecological and financial outcomes that you must weigh against your landowner goals. However, if we are to going to use nature as our template for forest management, we must consider forest time in our decisions.

Principle 4: Context

Forests and the processes that shape them know no boundaries. Your forest establishes, grows, and develops at a scale far greater than the size of your property and as part of the forests to which it is connected within your landscape. While you can make decisions only for your own land, that does not mean that the perspective you take can't operate at a larger scale.

While standard forestry often focuses on specific sections of your forest, ecological forestry encourages decisions that take into account the interconnectedness of your forest to all around it. By shifting your decision-making to the landscape level, you can see the larger picture, which better represents the way in which your forest functions. The smaller your property, the more important it is to consider the context of your forest, as it is even more dependent on the forests around it to sustain itself and provide benefits.

Focusing on the context of your forest should also include the social landscape, including opportunities to share knowledge and experience with your neighbors. The full realization of applying the context principle to your property is the actual coordination of forest management with neighbors. This could be as informal as timing your creation of wildlife habitat with your neighbors to achieve greater impact, or something as formal as having a joint timber sale.

Only by understanding the landscape context of your property can you steward your forest most effectively.

FORESTRY BASED ON RELATIONSHIPS

Ecological forestry shouldn't be thought of as a cookie-cutter approach that can be applied on every property in the same way. In fact, it's the opposite. At its fullest expression, forestry is both science and art. The science component of forestry is critical to expanding our understanding of forests. The art component of forestry lies in applying this knowledge to your unique forest, embedded within your unique landscapes, and guided by your unique combination of landowner goals.

The practice of ecological forestry exists along a continuum. The knobs can be turned to low, medium, or high, and, of course, all points in between. You can refine standard forest management a little or a lot. The degree to which you refine it should be guided by your combination of landowner goals. Importantly, no matter how much emphasis you place on certain goals, there is always a way to practice ecological forestry by implementing the guiding principles. Even seemingly small efforts to incorporate the principles of ecological forestry within your forest management plan can have large impacts on your forest and within the landscape, especially when added to the efforts of your neighbors.

There is no better way to fully understand the principles of ecological forestry than to apply them to your own forest. Seeing firsthand the ways in which forests work and respond to your stewardship practices provides insights that develop judgment, a critical aspect of all decision-making. Importantly, practice implies humility. If you were perfect, there would be no need to practice. There may be mistakes, but that's okay. There are mistakes in all relationships. Learn from them and incorporate those lessons into the next decision you make about your forest. Ecological forestry is meant to be practiced humbly.

Know that you are not alone. You don't need to become an expert. Take the pressure off yourself. There are trained, experienced foresters working in your community that can help you understand your forest and its context, help you clearly define your goals, and make recommendations on strategies to implement these principles of ecological forestry in your forest. Chapter 5 provides information to help you find a qualified forester who is right for you.

The practice of ecological forestry is about creating a healthy relationship with the forest and our neighbors. As such, your role is more than simply adding to the number of acres of land that are managed with ecological forestry. The ways in which you choose to engage with forests and with your neighbors result in more than increased deadwood and old trees or the limitation of the spread of invasive plants, though these are very good ecological outcomes. Your adoption of ecological forestry helps to create a shift in culture—a shift in our relationship to forests. Rather than considering ourselves as separate from forests, it recognizes that people are active participants within the forest and that, as such, we must find balance between our needs and those of forests.

Ecological forestry is the implementation of a reciprocal relationship between forests and people. It expresses that ultimately what is good for forests is good for humans. It expresses the land ethic that Aldo Leopold wrote about and helps to achieve a relationship of reciprocity as written about by Robin Wall Kimmerer and many other Indigenous people for centuries before Leopold. It is this reciprocal relationship that will ultimately sustain both forests and people into the future. You can see it in the stories of landowners across the region, including those who are featured in the case studies within this book: Alan Finifrock in Minnesota, Ed Neuhauser in New York, Tim Stout in Vermont, Mason and Ina Phelps and the Riley family from Massachusetts, Sally Hightower in Michigan, and Janet Sredy in Pennsylvania. The spark of this relationship exists. We need it to grow and spread.

In many ways, ecological forestry can be summed up as the recommendation to focus on creating a healthy and reciprocal relationship with forests and people. If you're focused on those relationships, good things will follow.

Practicing ecological forestry is about creating a healthy relationship with your forest by engaging with it regularly in ways that meet your needs as well as those of the forest.

PART TWO
Practicing Ecological Forestry

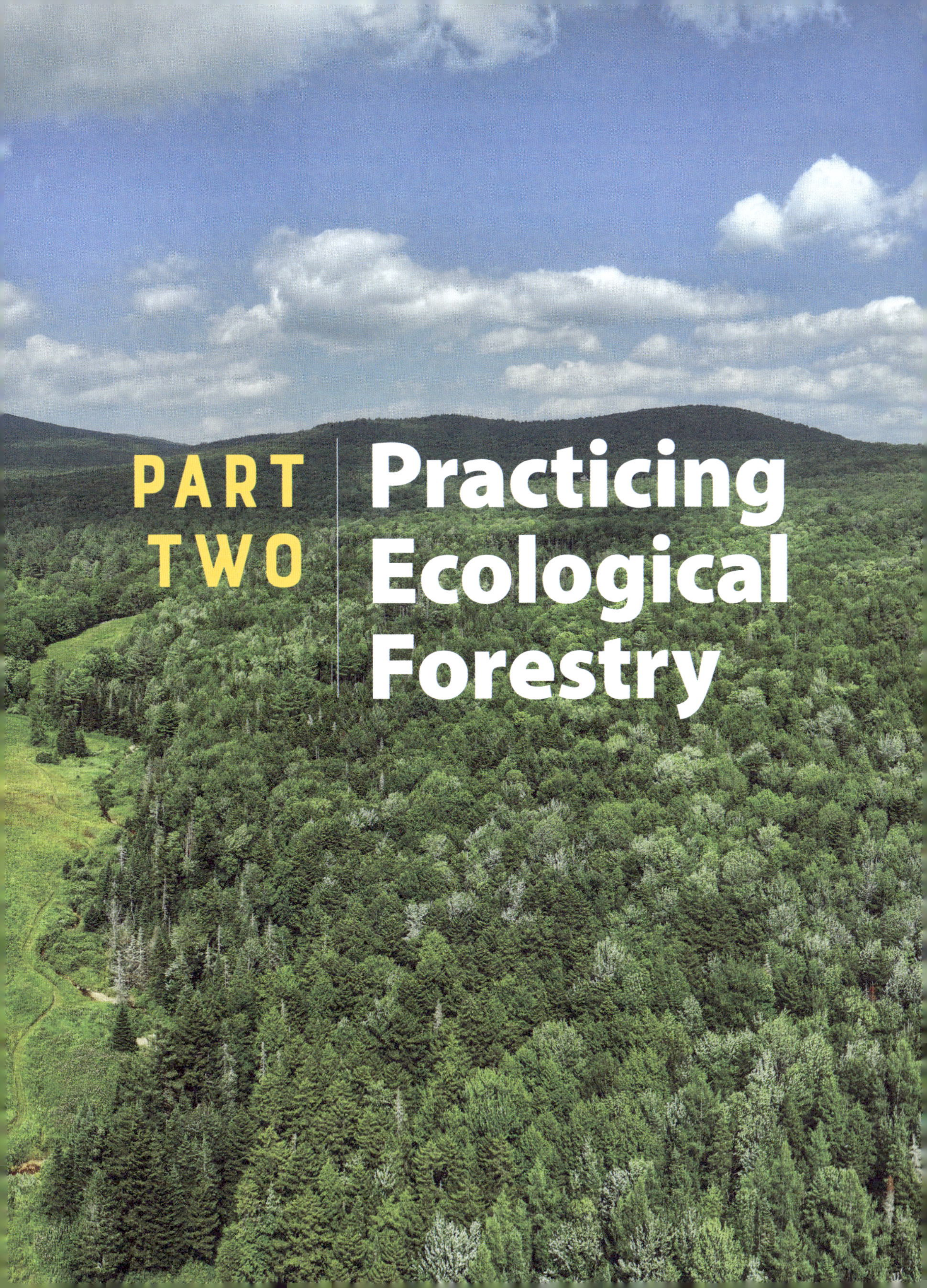

“To both use and conserve nature requires complex knowledge and practices, far more complex than leaving nature alone.”

—M. Kat Anderson, *Tending the Wild*

CHAPTER 5

Choosing Your Stewardship Approach

AS A SOCIETY, WE CONTINUE TO EXPECT MORE and more from our forests, yet the challenges they face seem to increase each year. What will our response be in this defining moment? What will be our legacy for the future? As a forest landowner, the choice is yours and that of thousands of your neighbors. As a collective, the decisions you make about the stewardship of your forests will have a profound influence on the forests of the future and the benefits they provide (or don't!). It's your turn to shape our forests and our relationship to them.

If this seems overwhelming, don't worry: You're not alone. This chapter will introduce you to the basic differences of active versus passive management and the role that a forester can play in helping you define and realize your goals.

Your decisions (and non-decisions) about the stewardship of your forests have impacts for you and the forest. Foresters are available to help inform these decisions.

CONSIDERING THE MANY BENEFITS OF FORESTS

As you begin to consider how best to steward your forest, you may want to pause and think about why your forest is important to you as a landowner. Your forest may provide a beautiful and private backdrop to your life, a place to recreate, and the opportunity to view wildlife. Your children may have built forts, caught salamanders, and explored the wonder of nature within your forest. Timber harvesting may even provide you with periodic income to help you afford property taxes.

Beyond the joy and personal benefits you receive from forests are the many important benefits of forests to society. Humans have always relied on forests. Some of those benefits have remained the same over thousands of years: fuel for heat, building materials for shelter, clean water to drink, a place for spiritual renewal. Other benefits, such as climate change mitigation, are ones our ancestors could never have imagined, just as future generations will likely rely on benefits that we haven't considered. One thing is certain. The benefits our forests provide are life-sustaining now and will always be so.

While owning forests comes with many benefits, it also comes with responsibility. Sustaining the many personal benefits that forests provide, as well as the benefits on which society depends, while also helping to sustain forests, requires making informed decisions about their stewardship.

DEFINING YOUR GOALS

Start by taking the pressure off yourself. Like many landowners, you want to be a good steward of the forest and do the "right" thing by it, but there isn't one "right" strategy to forest stewardship. Simply keeping your land in forest use sustains many of its benefits and keeps your forest management options open.

Forests provide many benefits to their owners and society, including wildlife habitat, wood products, firewood, recreational opportunities, climate change mitigation, and clean water. As a landowner, you can choose the specific benefits that you would like to encourage from your forest. These become the goals that your forest management strategies will strive to achieve and sustain.

Most landowners have a combination of ecological, personal, and financial goals they are working toward. The more specific you can be about your goals—and about the order of their importance—the better. For example, many landowners rate wildlife habitat as a high-priority goal. However, different types of wildlife use different stages of forests. If you are able to be specific about the type of wildlife you're interested in supporting, it can help inform your forest management strategies. Like everything in life, there are times when trade-offs must be made. Having a clear sense of your priorities makes decisions about these trade-offs much easier. You will also need to decide the extent to which you would like to adopt the principles of ecological forestry. Ecological forestry exists on a gradient, and your adoption of ecological forestry should be tailored to fit your goals.

Your goal setting should also include plans for the future of your land. Who will own your land in the future? How do you want to see it used? Defining the goals for your land, particularly those about its future, often necessitate conversations with spouses, children, or others who may be involved with the land. Of course, the constellation of people whom you decide to pull into the decision is entirely up to you.

When defining your goals, be sure to consider the specific characteristics of your land and the surrounding landscape. Doing so may provide forest management strategies that you had not considered. You may find out that your land is an important connector between forest blocks and maintaining the connection is critical for species migration. Or you may find that your forest management can help create and maintain important habitat lacking in your landscape. Likewise, you may also discover challenges that make meeting certain goals difficult or even unrealistic. Your landscape may have a new invasive insect or plant to which your forest is vulnerable.

When taken together, your unique combination of goals and your land's own characteristics will lead to a unique set of forest management strategies. The strategy that is a good fit for one landowner and their forest is not necessarily a good fit for another. Never forget that this is your land and the management implemented on it should match your goals. Don't get talked into something you don't understand or don't feel comfortable implementing.

WORKING WITH A FORESTER

Defining your goals and making decisions about your forest can seem overwhelming. There is a lot of information to consider, and stewarding your forest likely isn't your full-time job. A lack of time and information can lead to indecision and inaction. The first piece of good news is that forests know what to do without you, so you have time to make sure you have the information you need to feel comfortable making a decision. The other piece of good news is that you don't have to steward your forest alone. There is an entire cadre of foresters and other professionals working locally to help you understand your options and apply strategies that will meet your goals.

A forester is a professional with a college-level education and experience in a broad range of forest-related areas, including ecology, soils, forest health, wildlife management, geographic information systems (GIS), forest management, forest policy, forest economics, and the growing and harvesting of forest products. An increasing number of states are enacting forester licensing programs that require professional foresters to apply for a license, which is often based on the person's education, knowledge, and experience. Once licensed, foresters are then required to maintain their license through continuing education and professional conduct.

Working with a forester to assess your forest and its landscape context and define your goals will help clarify the actions you can take to sustain your forest and its many benefits.

THE DIFFERENCE BETWEEN A FORESTER AND A LOGGER

Landowners sometimes confuse foresters and loggers, but they are two different professions, each with their own role in the forest. When considering the role of foresters and loggers, it can be helpful to make an analogy to the relationship between architects and builders. An architect working to design a house for a client must clearly understand the client's goals to meet their needs. Does the client want a single-floor home or two stories? How many bedrooms do they need? What is their budget for the building project? Once the plan is complete, a builder is then brought in to implement the blueprints of the architect and build the house. A well-built home requires both a thoughtful plan from the architect and expert implementation from the builder. Without competent professionals of both types, the house project will not be successful.

Similarly, a forester works with landowners to understand their forest and their goals in order to design forest management strategies to meet their needs. Is the landowner interested in wildlife and, if so, what kind? Does the landowner have income needs? Are the landowners planning to pass on the land to their children? A logger, an expert in felling trees and removing them safely from the forest, is then brought in to implement the strategies designed by the forester that require timber harvesting. A successful timber harvest requires a thoughtful plan from the forester expertly implemented by a logger. It takes both to have a successful timber harvest.

Types of Foresters

When planning your forest management, be sure you are working with an actual forester. If your state has forester licensing, ask the person if they are a licensed forester. If your state doesn't have forester licensing, ask them about their education and experience. There are also professional organizations for foresters, such as the Association of Consulting Foresters (ACF), Forest Stewards Guild, and the Society of American Foresters (SAF), that they may belong to. There are several different types of foresters that you may interact with as a forest owner.

Service and Extension Foresters

A service forester is hired by the state bureau of forestry to provide education and technical assistance to landowners. They typically work in specific districts and can provide you with general information and resources about your forest and your options to steward it. Service foresters may even be available to come out to your property and walk your woods with you. Since they are public employees, their services are already paid for through your tax dollars.

Extension foresters typically work for your state's land grant university and provide research-based educational programs and resources to help family forest owners make informed decisions about the stewardship of their forests. Different states have different configurations of service and extension foresters. They are both an excellent first call for landowners starting to investigate their forest management options. For more experienced landowners, establishing a relationship with these foresters can help you stay connected to new information and opportunities. While they are a great source of information, service and extension foresters typically do not implement on-the-ground forest management. For that, you will need to hire a consulting forester.

Consulting Foresters

A consulting forester is a trained professional paid directly by landowners to provide forest-related services. Commonly offered services provided by consulting foresters include conducting forest inventories, writing forest management plans, assisting enrollees in current use tax programs, conducting timber harvests on behalf of landowners, controlling invasive plants, conducting wildlife habitat projects, implementing prescribed burns, marking property boundary lines, and making maps of the property. Consulting foresters are "boots on the ground" professionals who can help you define your goals and then develop forest management strategies to meet them.

A consulting forester is a partner in your forest management, acting as your agent. As when choosing any professional, foundational to your relationship with a consulting forester is the ability to communicate with one another openly and honestly. If your consulting forester doesn't understand your goals, they will not be able to implement strategies that meet your needs. Yes, the consulting forester is the expert in your relationship, but it is your land. Find someone you are comfortable with and who shares your ethics about the land and understands you and what you're trying to achieve. The aesthetic, financial, and ecological implications of your forest management decisions can last decades. Getting it right requires effective communication and trust.

The importance of finding the right consulting forester can't be overstated. It is well worth spending your limited time and energy to identify the right one. Once you have the right consulting forester, many things fall into place, such as forest management that matches your vision for the forest, learning about opportunities that can assist you in reaching your goals, and connecting with other professionals within the consulting forester's network who are equally good, such as loggers, wildlife biologists, or someone to control invasive plants. Below are some considerations when choosing a consulting forester:

- **Consider the forester's location.** Most consulting foresters will work within a radius around their home or office. Having a forester who lives close to your land can be an advantage if it comes time to implement forest management strategies that benefit from their supervision, such as a timber sale or wildlife habitat work. It will be easier for them to drop by to check on the job, ensuring the work is meeting your goals. A forester close by is also bound to have more local contacts.
- **Understand how the forester will charge you.** Consulting foresters charge for their forestry services in different ways. They may charge an hourly rate for boundary work, by the acre to write a forest management plan, and a percentage of the proceeds of a timber sale. Be sure to get an estimate of the total cost for the service(s) you're hiring them for.
- **Ask about the forester's specialties.** Although there are core activities that all consulting foresters can help you with, such as writing a forest management plan or setting up a timber sale, there are some consulting foresters who have specific areas of expertise. Some are particularly interested in bringing back preforest wildlife species. Others are highly engaged with invasive plant control. Similarly, some consulting foresters may have particular expertise in geographic information system (GIS) mapping, while others have experience laying out and installing recreational trails. When searching for a consulting forester, be sure to ask them about their specialties to find the best match for your own interests.

How to Find Your Forester

So, if finding the right consulting forester is so important, how do you go about it? To start, you can contact the local service or extension forester. As public employees, they typically cannot make personal recommendations, but they often have a list they can provide of local, qualified consulting foresters. Many landowners also find it very helpful to get a referral from a trusted source, which narrows down a general list to a more manageable number of foresters to consider. Finding a referral for a consulting forester requires tapping into your local social network of professionals and neighbors. A land trust is a nonprofit conservation organization that works with landowners interested in keeping some or all of their land in its natural state. Your local land trust may be a source of information about consulting foresters, and since land trusts are not public agencies, they can make personal recommendations. Contacting your local land trust has the added benefit of starting a relationship, which could help as you begin your conservation-based estate planning.

Of course, the best way to find a consulting forester that's right for you is to talk to friends and neighbors with experience in forest management. Talking with a peer provides the opportunity to learn from the experiences of someone like you. Finding out about both their good and bad experiences with foresters and forest management can be equally valuable to your decision-making. Whom did they work with? Was it a good experience? Was the person responsive to their needs? How do they charge? Would they hire them again? What would they do differently? If you don't have any friends with forest management experience, consider contacting your county or state forest landowner association or attending a local extension event. As when choosing any professional, it is always helpful to speak with several consulting foresters to get a sense of your options. Putting time into finding the right consulting forester will pay off in the long run.

PASSIVE AND ACTIVE APPROACHES TO FOREST MANAGEMENT

It is essential to know the full range of forest management options, as this will help you better discern which combination of strategies best fits your goals, forest, and landscape. At the broadest level, forest management approaches can be described as either active or passive. However, let's not assume you need to choose one or the other. In fact, within the framework of ecological forestry, active and passive approaches are complementary parts of a larger whole, each adding important and different benefits.

Passive Forest Management

A passive approach to management allows natural disturbances to be the primary driver of change in your forest. The forest grows and changes on its own, with little or no human involvement. The passive approach shouldn't be thought of as a way to keep your forest as it is at this moment in time. Forests are dynamic and constantly changing. The forest will express itself through the ecological processes we've described and in response to the forest disturbances that act on it. The forest succession and development clock will move forward and backward as species composition shifts. The forest will move through the stages of forest development. If your forest is currently meeting your goals and will continue to do so in the foreseeable future as it changes on its own, then the passive approach could be a good fit for you.

Of course, the devil is in the details when determining what "little or no human involvement" actually means. At the most extreme end of the passive approach there are those who advocate for forested areas with absolutely no human intervention, including no passive recreation. However, for most, the passive forest management approach includes some form of forest-human interaction, which could include nonmotorized recreation (such as bird-watching, hiking, and cross-country skiing), invasive plant control, hunting, and the collection of non-timber forest products for personal use (maple sap, branches for wreaths, mushrooms, etc.). While these activities involve engagement with the forest, they don't overshadow the

Passive forest management can include activities such as passive recreation, invasive plant control, and sugaring.

influence of natural disturbances and forest development.

Activities viewed as inconsistent with passive forest management have a disproportionate impact on forests relative to that of natural disturbances. Activities that are generally not consistent with passive forest management include harvesting timber, creating wildlife habitat, managing invasive insects, and producing firewood. The use of motorized vehicles is also generally considered inconsistent with the passive approach.

For some landowners, passive forest management fits their goals and preferred approach to forest management. However, like most things in life, there are shades of gray. Maybe the idea of a passive approach resonates with you, but you also want the option to heat your home with firewood from your own land. You could implement the above description of passive forest management but harvest a couple of cords of firewood to heat your home. Similarly, passive management may sound very good to you right now, but you also want the option to be able to react to an invasive insect in the future in a way that would involve harvesting at-risk trees to slow the spread. You could make that decision if the need ever arises.

The balance of uses consistent or inconsistent with the passive approach is for you to determine. As a guiding principle, the main goal of the passive approach is to allow natural processes to shape the forest. The more uses you implement that are described above as inconsistent with the passive approach (that is, the more human intervention), the less your forest's future will be driven by those natural processes. At some point, your activities tip the balance from a passive approach to an active one.

Benefits of Passive Forest Management

One very important function of passively managed forests is to serve as a benchmark. We are facing an uncertain future. We must be humble and recognize that it is essential to have templates of nature's response without human intervention. Having passive management applied across landscapes at different scales will provide a control that can help us identify the differences between a hands-off approach and human intervention. Typically, public ownerships and conservation organizations with acreages far beyond that of family forest owners are excellent places to apply passive forest management for this purpose, since the large acreage allows natural processes to fully play out across the landscape. That's not to say that private landowners don't have an important complementary role to play in the landscape, as having smaller areas serving as benchmarks will also be informative.

The passive approach is one important tool in conserving biodiversity for those species that use old forest habitat for their survival. There have been a number of calls for the creation of large reserve areas across the globe to help stem the crash of biodiversity. Although public lands and those owned by conservation organizations are the most appropriate places for these reserves due to their large sizes and ownership goals, it is widely acknowledged that these areas alone will not be enough to sustain biodiversity. Having smaller reserve areas scattered among the large reserves will complement them by providing stepping stones across the landscape so that species can move from one to another. Since most of the forests in our region are family forests, this means that you can play an important role by helping to support this habitat.

Finally, the passive approach will likely result in the greatest amount of carbon storage within your forest. As we will discuss in Chapter 7, forests typically become more complex with age, containing larger trees and an increased amount of deadwood. Since trees aren't removed from the site through timber harvesting, this often translates into more carbon stored in the forest.

What Will the Passive Approach Lead To?

For some landowners, the goal of passive forest management is the eventual development of old forests. If this is a reason for your adoption of passive management, it will be important to understand the challenges within your forest and within the landscape that will shape the forest's development. Unfortunately, given the challenges our forests face, we cannot assume that all forests will develop into an old forest condition as we currently think of it. Challenges such as invasive plants and insects and excessive herbivory can alter or even stop the successional trajectory of the forest so that it doesn't hit the threshold of old forest characteristics.

For example, if deer herbivory and invasive plants limit tree regeneration, then through time the forest becomes structurally simplified with combinations of species that are novel (relatively new to our forests), including understories of invasive species and not trees. The type and number of benefits that such forests provide may not completely meet the needs of their landowners or the public. It could also be that the centuries it may take for forests to respond to these challenges exceeds the time frame of your goals, in which case there are other strategies to consider.

For those landowners committed to the passive approach to forest management regardless of the future condition of the forest, the type and number of challenges facing the property may not matter. As we have noted, the forests of our region have shown themselves to be remarkably resilient, rebounding from agricultural clearing, centuries of natural disturbances, and even invasive insects and diseases. We have every reason to believe that forests will survive these current challenges. It may take centuries for forests to eventually adapt, but adapt they will. The specific species composition and structure of future forests will likely be changed, maybe dramatically, but there will be forest. Taking the passive approach, come what may, means having faith that ultimately forests will find a way through and around their challenges, leading to a new equilibrium.

Intentionally Choosing the Passive Approach

There is a difference between doing nothing in your forest for a number of years (perhaps because you've been too busy to do otherwise) and making an intentional decision to adopt the passive approach. Practicing passive forest management should be intentional. Before choosing any management strategy, consider the full spectrum of options and how the combination of approaches can help you reach your goals. Be clear about what area(s) within your forest will be passively managed and the specific uses you will implement, and communicate these to those who are involved with the land.

Importantly, a passive approach does not mean "do nothing." Even if you are not actively working to change your forest's structure or species composition through passive forest management, you can still engage with your property. Learning more about your forest as an ecosystem, monitoring for invasive species and pests, foraging mushrooms, taking a hike, or joining a network of other landowners are all ways to enjoy your woods. Given the current age of most of our forests, the development of old forests will take a century or more. Planning your woodland's future through conservation-based estate planning is critical for maintaining its benefits over time by ensuring it will remain a forest.

CASE STUDY

Whetstone Wood Wildlife Sanctuary

Mason and Ina Phelps

LOCATION: Wendell, Massachusetts

ACRES: 1,000+

When Mason and Ina Phelps first lived in Concord, Massachusetts, they loved being awakened on summer mornings by the song of whippoorwills. However, after a few years, the abutting land was sold and four houses were built on the newly created cul-de-sac, Whippoorwill Lane. Ironically, Whippoorwill Lane took away the whippoorwill habitat. The Phelps no longer heard the song of the whippoorwill. They realized that the conservation areas that they enjoyed existed only because a private individual made the decision to conserve their land. Out of this loss and realization, an idea was born. Mason and Ina would buy a piece of land to provide a sanctuary for wildlife.

Soon after starting their search for land in Concord, they realized that the parcels of land were too small and too expensive to meet their goal of a wildlife sanctuary. They extended their search to central Massachusetts and, in 1961, bought 235 acres of land in the town of Wendell. Pleased with their purchase, Mason and Ina began working to "improve" the sanctuary through plantings of wildlife-friendly plants and encouraging native wildflowers.

The Phelps were soon approached by an abutter to find out if they would be interested in buying his land. They decided they were interested in adding land to their sanctuary and purchased the property. This experience made them realize that it was possible to further expand their sanctuary, so they revised their goal and began to work toward acquiring as much land as possible to protect Whetstone Brook, which flowed through their property.

Over the next few years, the Phelpses contacted more than twenty landowners and eventually acquired more than 1,000 acres of land, a mosaic of forests, wetlands, and streams hosting a diversity of wildlife. They named the sanctuary Whetstone Wood. The Phelpses then worked with several conservation organizations, eventually teaming with Mass Audubon to significantly expand and permanently protect Whetstone Wood, creating the sanctuary they had long envisioned.

The Phelpses' philosophy of land management changed over their decades of ownership. Their original vision of the land in the 1960s was to create a pastoral landscape. By the 1970s, they had come to view their land as a wildlife management area in which the habitat could

A combination of passive and active approaches enhances the ecological function and biodiversity of the land as a whole.

be manipulated through plantings, forest management, and mowing to create varied environments and encourage diverse wildlife populations. Eventually, after watching the forest grow and change on its own, the Phelpses made the decision that the best way to manage Whetstone Wood was to take a passive approach and let nature take its course.

Additional land conservation efforts of their neighbors have resulted in the Whetstone Wood growing to more than 2,000 acres. Whetstone Wood has been sold to Mass Audubon and has the distinction of being their largest wildlife sanctuary overall. Whetstone Wood serves as a passively managed core habitat embedded within approximately 50,000 acres of conservation land that employs active forest management strategies to meet landowner goals, including a state forest, a state wildlife management area, drinking water supply, and lands owned by a local land trust. The passive approach of Whetstone Wood and the active approach of the surrounding conservation lands complement one another, enhancing the ecological function and biodiversity of both areas.

The Phelpses leave a tremendous legacy to wildlife and to their community. They saw a need and worked diligently over decades to be the agents of the change they sought. And to think it all started with the call of the whippoorwill.

Active Forest Management

Active forest management involves intentionally shaping forest composition and structure to produce desired forest benefits. Due to our region's land use history, many of our forests are dominated by trees of the same age, approximately sixty to one hundred years old. This uniformity of our forests has created vulnerabilities that can interfere with a variety of common landowner goals. Active forest management affords us the opportunity to diversify our forests and the benefits they provide, including wildlife habitat, resilience, and wood production.

Most wildlife species rely on a mix of forest stages, including preforest and old forest stages. In addition, there are a number of species in decline that will only use the preforest stage. Active forest management can diversify forest development stages to improve wildlife habitat for the full suite of native species. We can turn the successional clock back to create forests in the preforest stage, and also speed up the clock and reduce the time it takes for our forests to obtain old forest structure.

Having forests of similar tree species and development stages means we have a lot of our "eggs in one basket." Using active forest management to diversify species and developmental stages reduces vulnerabilities by spreading out risk of impact from a disturbance and increasing forest resilience. Increasing resilience within our forests can help maintain desired benefits, including avoiding significant losses of stored carbon due to large-scale disturbance.

Finally, our society is a tremendous consumer of wood products. With active forest management we can meet our own wood needs and provide economic benefits to people and communities by producing it locally.

While these topics will be covered in greater detail in the following chapters, they help to demonstrate the opportunities to reach your goals through active forest management. Active forest management gives you the chance to intentionally turn the knobs up and down on forest benefits to meet your goals and those of society.

A Deeper Relationship

While there are many tangible benefits of active forest management, there are also some intangible benefits that are not often discussed but are critical to articulate. A deeper relationship with the forest will grow out of thoughtful, respectful, and humble active forest management. Kat Anderson's quote from *Tending the Wild* at the beginning of this chapter nails it: "To both use and conserve nature requires complex knowledge and practices, far more complex than leaving nature alone."

The complex knowledge that we need in order to both use and conserve forests can only be acquired through some form of active engagement with them. Forest stewardship is the pathway to an intimate knowledge of forests. Intimate knowledge of forests provides insight into ways in which we can responsibly enjoy the life-sustaining benefits of forests while also reciprocating to ensure forests have what they need to sustain themselves.

Our culture has lost its connection to, and appreciation of, forests. Few know the names of trees, the way forests work, the creatures that live there, and the ways in which we rely on them. Forests have become an abstract idea instead of a part of our daily lives. Society celebrates and even romanticizes farmers and their relationship to the land, which is born out of working it, but doesn't value that same relationship between landowners and their forests. Practicing ecological forestry through active forest management can help restore the connection we have lost. It can stimulate a shift in culture, for in practicing ecological forestry, we, too, are changed. As Aldo Leopold reminds us, "That land yields a cultural harvest is a fact that is long known but latterly forgotten." The need for a cultural shift within our society and the deep connection that active forest management cultivates provide compelling rationale for active management—beyond the tangible benefits of this approach.

ACTIVE OR PASSIVE STEWARDSHIP: WHICH IS RIGHT FOR YOU?

Gather information about your woodland and determine what stage of succession it is in and what benefits it currently provides. Try to articulate your goals for the property based on your use and needs.

CASE STUDY

Starting Active Forest Management

Sally Hightower

LOCATION: Gladwin, Michigan

ACRES: 40

After growing up in the city, Sally Hightower was eager to stop renting and to purchase land for herself when she moved to rural Michigan for a teaching position. Through connections at the school, she was able to get a deal on a 40-acre piece of property outside of town. Sally and her wife put in a 600-foot driveway and built their house on the beautiful banks of the Tittabawassee River.

Feeling a drive to improve the land, Sally has put in a lot of effort over the years. Her first step was to hire a consulting forester. He helped them plan a timber harvest and connected them with a responsible logger to implement it. The forester also pointed them in the direction of the American Tree Farm System, which he thought would be a great fit for their land.

Fostering community and connection has helped bring Sally closer to her goals.

Sally cares deeply for her woodlands and is saddened when trees are harvested or removed. However, she recognizes the benefits of harvesting. The property has had multiple harvests for pulp and pallet wood, timber stand improvement, and wildlife-specific management. She even convinced a couple of college students to plant 1,200 red pine seedlings for her over the course of one summer.

In the years since she purchased the property, the surrounding parcels have been split up into smaller acreages, many of which are vacation spots for people coming up from the cities. In the future, she wants to maintain the forest's healthy condition but sees it as remaining mostly unchanged, so she worked with her local land trust, the Little Forks Conservancy, to donate a conservation easement.

Sally works with her neighbors to help inform them and try to get them engaged with their woods. She would love to see the green signs of the American Tree Farm System all up and down the road. Fostering community and connection might be valuable for her management goals, too. She has a 10-acre stand ready to be harvested but struggles to find a logger to do a job this small, so she is contacting neighboring landowners to come together for a joint harvest. She also communicates with the anglers who use her land and works hard to control erosion and manage trout habitat.

Sally received the 2018 Michigan Outstanding Tree Farmer of the Year award, yet with all her experience, she still feels there is much to learn about her forest. Her love for her land keeps her going. During our interview, she shared, "I would have never imagined this city kid living in the woods and owning a tree farm and doing management practices, would have never thought it possible in my wildest dreams, but I am so thankful that I live here. And the first time I stood on this ridge where my house is and looked down at the river, I just had a peace come over me. I just felt total contentment and peace. I said, this is where I'm supposed to be, and I have never looked back. I love it out here."

ADOPTING BOTH PASSIVE AND ACTIVE MANAGEMENT

Although passive and active forest management sit at opposite ends of the spectrum, it would be a mistake to think of them as conflicting strategies. There are those who feel very strongly that the passive approach is the best way to steward forests. They believe that, to benefit wildlife and mitigate climate change, trees should not be cut down. There are those who feel just as strongly that the active approach is the best way to steward forests. They believe that trees *should* be cut down—to benefit wildlife and mitigate climate change and to meet society's need for wood products.

Consider the possibility that both are correct. Each approach provides different but necessary benefits. We need both approaches out on the landscape. It all depends on the specific goals and the landscape context. We look to our forests to provide a wide diversity of benefits. This diversity of benefits comes from a wide diversity of strategies, ranging from the passive approach to intensive forest management and everything in between. Therefore, passive and active forest management are not conflicting strategies but rather complementary strategies within the larger framework of ecological forestry.

While you may decide to implement either the passive or the active approach throughout your entire property, you can also decide to do both. You may decide to take the passive approach on one part of your land and pursue active management on others. If that is of interest, then consider passive forest management in areas of your property with high ecological value (such as wetlands, stream sides, and endangered species habitat) or cultural value (for example, precolonial stone structures or unique colonial-era features). Also consider areas on your property with relatively few challenges, such as invasive plants, so that your forest can function normally and develop into an old forest.

To maximize the value of passive areas, it is ideal to locate them in diverse areas—wet and dry, high and low elevation—and in different forest types. The passive approach can be implemented in large areas or in patch reserves of less than an acre. At its smallest scale, you can implement the passive approach at the tree level, by identifying individual trees that will never be harvested, but instead maintained so that they may grow large and old, living out the natural course of their lives.

Areas of your property that may make sense for active forest management would include stands that have forest challenges in need of addressing, such as high concentrations of trees vulnerable to an invasive insect, as well as areas without sensitive soils, with good access for forestry machinery and public roads, with limited special ecological resources, and with good soil fertility. When evaluating the property, it may make sense to first identify the areas for the passive approach and then consider the balance for active management.

Don't forget to consider the landscape context of your land. Strive to ensure that there are both passive and active approaches within the larger landscape. Remember that, historically, the incomplete nature of natural disturbances meant that some forests within the landscape would have been recently disturbed while most wouldn't have had a large disturbance for centuries, resulting in part of the landscape being in the preforest stage and the majority being old forest. A combination of passive and active forest management approaches within the landscape can help us mimic this condition in our current forests.

PUTTING IT INTO PRACTICE

With a set of well-defined goals and a consulting forester with a clear understanding of your vision for your forest, it's time to begin. The following chapters provide detailed information on some of the most common landowner goals and critical issues facing our forests. Whether you are leaning toward the passive approach, the active approach, or both, each chapter provides opportunities to learn about strategies that you can adopt to practice ecological forestry in your own forest.

Importantly, it is not the intent of this book to turn you into a forester. You don't need to be an expert. Instead, the goal is to help inform your decisions about your forest by increasing your confidence to effectively communicate with a forester and move forward. Practicing ecological forestry can take many forms. The hardest step is often the first one. Start your practice of ecological forestry by choosing a strategy that interests and excites you. Let your relationship with the forest grow from there.

> “To keep every cog and wheel is the first precaution of intelligent tinkering.”
>
> —Aldo Leopold, *Round River*

CHAPTER 6

Restoring Old-Growth Characteristics

PERHAPS NO FOREST MANAGEMENT GOAL DEMONSTRATES the application of ecological forestry more directly than restoration of old-growth characteristics to the ecologically young forests of our region. Forests dominated up to 90 percent of the region at the time of European settlement. In the nineteenth and early twentieth centuries, widespread clearing reduced forest cover to roughly 30 percent. Areas not converted to agriculture were greatly simplified by extensive forest harvesting.

The last century and a half have witnessed an incredible rebound of our forests. Even though our forests have grown back, there are few remaining examples of old-growth forests, which now occupy less than 1 percent of the region. While we cannot recreate true old-growth forests, we have opportunities through both passive and active forest management to restore old-growth characteristics to our ecologically young forests.

Old trees and old forests have unique characteristics.

WHAT IS OLD GROWTH?

Old-growth describes forests that were never directly affected by intensive human land use, such as that implemented by Europeans. Old growth is a very rare condition in the regions covered by this book due to the enormous impacts of agriculture, logging, and other land uses beginning in the colonial era. Cultural practices, such as the use of fire by Indigenous peoples, certainly influenced the condition of historic forests, particularly near villages and other settlements, but they are considered harmonious with natural processes.

Second-growth forests established and grew following intensive human land use, such as agriculture or logging. Most of the forests currently found in our region are best described as second growth.

Old-forest describes a forest that contains a critical mass of characteristics associated with old-growth forests: large and old trees, spatial variation in tree density and size, abundant deadwood, and multiple canopy layers. Although specific forest ages have been included in definitions of old forest, the ecological time frames over which old-growth characteristics develop vary considerably by forest type, disturbance history, and site fertility. As a result, conservation strategies should focus on restoring characteristics associated with old growth rather than relying solely on estimates of tree age.

WHY IS OLD GROWTH IMPORTANT?

Our current forests are much different from the forests our native plants and wildlife evolved with over thousands of years. Some plants, lichens, mosses, fungi, and invertebrate species are dependent on old-growth characteristics that are currently lacking or less abundant in our second-growth forests.

Also, many species—particularly native birds, including some woodpeckers, warblers, and thrushes, as well as certain mammals, such as fishers and martens—reach greater abundance in forests that have large trees with cavities and complex canopy structures. These abundant populations, called "source populations," are crucial for populating or repopulating surrounding habitats and are therefore central to the long-term viability of our native species.

In the context of climate change, it is widely recognized that although they do not sequester carbon as quickly as younger forests, old-growth forests store the greatest amount of carbon. Therefore, mitigating climate change requires increasing the number of forests with old-growth characteristics on some parts of the landscape while also encouraging a diversity of other forest age classes in other areas. Finally, old-growth forests support cultural traditions, providing human wellness benefits and the opportunity for spiritual renewal. These human values, in addition to the tangible ecosystem services these forests provide (water storage and filtration, flood resilience, localized cooling, climate buffering), underscore the importance of these forests to our well-being and even survival.

As we grapple with a changing climate and a proliferation of nonnative insects and diseases, there is great uncertainty regarding how forests will adapt. Encouraging more of the forest structure and species found in old-growth forests helps us keep every piece of the puzzle that our forests naturally evolved with, undoubtedly increasing their resilience.

OLD-GROWTH CHARACTERISTICS

Our region has a great diversity of forest types. Despite these differences in dominant species, there are common characteristics generally more abundant in old-growth forests, regardless of forest type. This section summarizes some of those key characteristics, which are often lacking in the second-growth forests you see today.

Presence of Large and Old Trees

The first characteristic that comes to mind when thinking about old-growth forests is big trees. It is common and understandable to relate a tree's size to its age. While, in general, smaller trees are younger and larger trees are older, that's not always the case. A tree's height is related to the site's fertility and available soil moisture: The more fertile the site, the taller the tree. A tree's diameter is more directly tied to how much space it has to grow in.

Nonetheless, a common characteristic used to differentiate old-growth forests from second-growth forests is the abundance of large-diameter trees. In old-growth northern hardwood forests, the standard is 12 to 20 large-diameter trees per acre, each with

Old-growth characteristics include large and old trees, ample standing and downed dead trees, and a variety of tree sizes and ages.

CHARACTERISTICS OF AN OLD-GROWTH FOREST

A. Gap in the forest created by a natural disturbance resulting in standing and downed deadwood. Increased light conditions will promote the release and establishment of young trees.

B. Large-diameter (20"+ DBH) standing dead tree (snag).

C. Large-diameter (20"+ DBH) overstory trees.

D. Seedlings and saplings from a previous gap create multiple canopy layers, variation in tree size and density, and species diversity.

a diameter at breast height (DBH) of at least 20 inches.

The lifespan of the long-lived tree species often found in old-growth forests (hemlock, sugar maple, red spruce, red and white oak, cedar, white and red pine) ranges from three hundred to more than four hundred years. By comparison, our approximately one-hundred-year-old forests are ecologically young, which accounts for their lack of old-growth characteristics.

Beyond their size, these old trees have developed unique bark and branch patterns, which support a diversity of native insects and provide perching habitats for birds that eat those insects. In addition, old trees often have cavities that serve as homes for a wide range of wildlife. In short, old trees show their age!

Spatial Variation in Tree Density and Tree Size

Although big trees are the first characteristic many associate with old-growth forests, a key attribute of these ecosystems is the mixed sizes, ages, and density of living and dead trees. Such variability is the result of natural disturbances such as wind, insects, and disease that kill individuals and groups of trees, creating a mosaic of canopy gaps and groups of different tree sizes and ages across the forest over time.

This mosaic was likely reinforced in our fire-dependent ecosystems (for example, pitch and jack pine barrens and oak and red pine woodlands) by historic prescribed burning applied by Indigenous peoples. Beavers, too, contributed greatly to the ever-changing landscape prior to their widespread extirpation.

Abundant Downed Deadwood in Various Sizes and Stages of Decay

Natural tree death is an important ecological process, both in generating the openings in the forest canopy necessary for new trees to establish and in creating deadwood, which serves as critical habitat for many species and contributes to carbon storage, moisture retention, and soil protection. It is estimated that old-growth forests often have at least two to four times the amount of downed deadwood found in second-growth forests.

Downed trees are the consequence of periodic disturbances through time. Trees that have fallen only recently are just beginning to decompose, and others that fell decades ago are well decomposed already. In addition, since there are often large trees in old-growth forests, the downed trees are usually large in diameter, which makes them more resistant to decay. These large-diameter downed trees are quite rare in our current forests. They provide habitat for organisms that break wood down, supply foraging opportunities for other species, and serve as "nurse logs" for tree species that regenerate well on decomposing wood, such as hemlock, red and white spruce, northern white cedar, and black and yellow birch.

Dead trees also contribute to the habitat of streams, wetlands, and vernal pools and provide key ecosystem functions, including storing and cycling nutrients and carbon and retaining moisture.

Large-Diameter Standing Dead Trees

Although less abundant than downed trees, large-diameter standing dead trees (snags) are also common characteristics of old-growth forests. Snags serve a crucial role in providing foraging opportunities for birds (such as woodpeckers) that excavate the deadwood in search of insects. The holes that are created in the wood then serve as habitats for various birds and small mammals, with larger snags even supporting large mammals such as fishers, martens, and black bears. Given that our current forests are lacking many large-diameter living trees because of their relatively young age, large snags are even less common.

Dead standing and downed trees are critical biological legacies that provide habitat to a number of species and store carbon.

Multiple Canopy Layers

The diffuse light and deep shade you often encounter when visiting an old-growth forest are a reminder of the many layers of leaves that sunlight filters through in these ecosystems. That filtering (and absorption) of light is the product of centuries of canopy development, which has allowed different species to grow to different heights beneath the main canopy over time. The resulting layers of foliage, often extending from the forest floor to treetops, creates a habitat-rich space that includes hiding and nesting cover close to the ground, as well as travel corridors for certain birds and other species that use the highly connected upper canopy layers. The multiple layers of vegetation also provide opportunities to store more carbon in the forest.

Regeneration

A key feature of the canopy layers found in old growth is the regeneration layer. This layer is typically dominated by seedlings and saplings of shade-tolerant species that thrive in low-light conditions—like hemlock, sugar maple, beech, red and white spruce, northern white cedar, and balsam fir—and have established themselves over time, often on well-decomposed logs or the forest floor. All forests rely on this preexisting layer of regeneration to recover following a disturbance. Having an established regeneration layer increases the resilience of the forest by ensuring that the next group of trees that will occupy the site is already in place when a natural disturbance kills the overstory trees.

Plant Communities

Few plant species are specific only to old-growth forest; however, these forests often have a greater abundance of some sensitive species, given that it takes a long time for them to recover in areas cleared by historic land use, particularly agriculture. Some of these species include forest herbs, such as wild sarsaparilla, dwarf ginseng, wood sorrel, squirrel corn, and cucumber root, as well as lichens (like lungwort) and feather flat moss. Many of these species maintain their populations through either asexual regeneration or dispersal of seeds by ants, making them quite sensitive to forest loss. Similar dynamics are also observed with many fungal species, which either require a long period of time to develop extensive belowground networks in symbiotic associations with maturing and old trees, or are reliant on the large, well-decomposed logs found in old-growth forests as substrates.

Well-developed understory plant communities are a sign of a forest that has been undisturbed for a long time.

ATYPICAL OLD GROWTH

We largely associate old-growth forests with forest types and locations on the landscape where large and old trees and closed-canopy forest conditions can develop. Nevertheless, there are many places, such as ridgetops and wetlands, where old-growth forests once existed. These areas are less likely to be actively managed to restore old-growth characteristics, but their protection through a passive approach is equally important to ensure that the full suite of old-growth forest communities once common across the region are restored and maintained. Examples of these "atypical" old-growth forests include dwarf pitch pine woodlands, cedar bluffs, black gum swamps, and black spruce bogs.

In addition to these atypical old-growth communities, other old-growth ecosystems are better characterized as woodlands or barrens, which naturally have far more open forest conditions than most of the forests predominating in our region. These more open forests are often associated with certain conditions (such as dry, nutrient-poor soils) and disturbance regimes (frequent, low-severity fires) that prevent the formation of dense forests.

Given that natural fires rarely occur at a high frequency in our climate, burning by Indigenous communities to encourage culturally significant wildlife and plant species was an important local factor in maintaining many of these ecosystems, including pitch pine barrens and oak woodlands and savannas. As such, those who wish to restore old-growth conditions in places where these communities once predominated would benefit from an understanding of traditional cultural practices to guide the application of prescribed fire and other management techniques.

MANAGING FOR OLD GROWTH

Old-growth characteristics can be restored to your woods through either passive or active management. Although there are landscape characteristics that may make one strategy more effective than the other, these approaches are highly complementary, and both are necessary across the landscape. In addition, the active management approach offers a gradient of strategies that can be used based on your interests in increasing the "old growthness" of your forests while meeting other goals (such as wildlife) and societal needs (such as locally produced wood products).

Passive Management

This approach involves letting nature take its course and waiting for forest development and natural disturbances to create old-growth characteristics without any direct human intervention. Designating an area of your land as a reserve without active forest management does not mean old-growth characteristics will appear in those places overnight. Developing these conditions with a passive approach will often take more than a century, based on the current age of most forests in our region, and is influenced by the fertility of the site, past agricultural impacts on the soil, and the amount of natural disturbance to which the forest is exposed.

In addition, the current condition of the forest will strongly influence the likelihood and development rate of old-growth characteristics. Dense forest plantations, abundant invasive plants, high deer densities, or beech bark disease may inhibit the development of key old-growth characteristics (large living and dead trees, multiple canopy layers) and inhibit or reduce biodiversity, carbon storage and sequestration, and resilience to climate change. As such, achieving old growth via passive management isn't a reasonable goal for every forest condition currently on the landscape.

The key drivers creating old-growth characteristics with a passive approach are natural disturbance events (such as windstorms, ice storms, and snowstorms) and insect and disease outbreaks. These events strongly influence forest structure and composition by creating gaps in the forest canopy, standing dead trees, pit-and-mound topography from blown-over trees, and downed logs. The progression and rate of change in passively managed areas are determined by the type, frequency, and intensity of natural disturbance. Since the most frequent disturbance events affect only single or small groups of trees, a passive approach will favor tree species that are most competitive in shade, such as hemlock, beech, sugar maple, balsam fir, and red and white spruce. Because no trees are harvested with this approach, it will produce the most natural appearance

PASSIVE PATHWAY TO OLD FORESTS

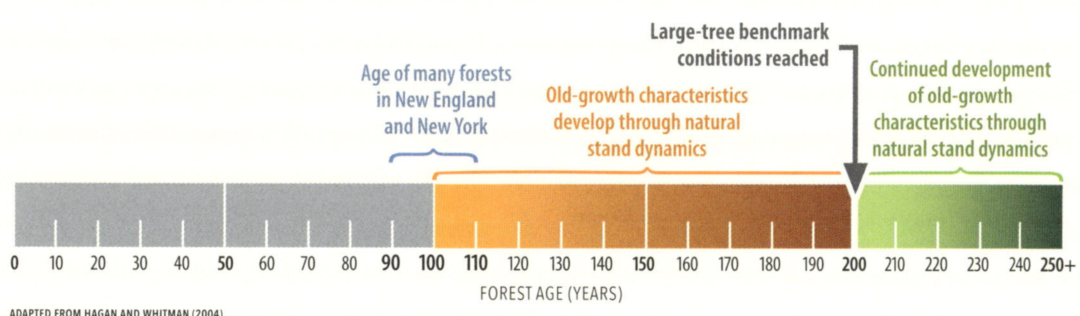

ADAPTED FROM HAGAN AND WHITMAN (2004)

SHOULD I SALVAGE MY WOODS?

Developing old-growth characteristics, whether through a hands-off or an active approach, involves leaving dead and dying trees in the woods. After a disturbance such as a windstorm or an insect or pathogen outbreak, landowners are often interested in salvaging trees that have been knocked down or killed. While it initially looks "messy," leaving these trees in the woods is critical to developing old-growth characteristics.

With both more frequent extreme weather events and encroaching invasive species (including the emerald ash borer, the spongy moth, the hemlock woolly adelgid, and the nematode that causes beech leaf disease), it is likely that the opportunity for salvage will only increase in the future. It is important to remember that these disturbances are generating key deadwood absent from most forests, canopy openings that allow for the establishment of regeneration, and complex young-forest conditions necessary for many species of wildlife. As such, limiting salvage operations or deliberately leaving some disturbed areas unharvested will ensure the positive legacy of these events by maintaining key structural conditions.

and conditions in your woods in the long run (no cut stumps or skid trails, and more deadwood remaining).

The passive approach to the restoration of old-growth characteristics will likely result in the greatest amount of carbon storage. As time goes on, trees get larger, more deadwood is added to the forest, and soils continue to increase as deadwood decomposes. The passive approach is therefore a good strategy if you are interested in storing carbon that has already been removed from the atmosphere; however, you should recognize that forests dominated by low-vigor trees and invasive species may not accrue these carbon storage benefits at a level similar to natural and healthy stands.

Landowners interact with their land in a variety of ways. The ultimate goal of the passive approach is to let natural forest processes and disturbances develop the missing characteristics over time. Activities that allow these processes to be the primary driver include creating hiking trails, controlling invasive species, removing a tree infested with an invasive insect, and even removing a few cords of wood per year for firewood.

Taking a passive approach to restoring old-growth characteristics will help meet the goals for increasing wildlands across the region. Since the time frame for developing old-growth characteristics is beyond that of a family forest owner, engaging in forest and estate planning (see Chapter 12) is critical to achieving and sustaining old-growth characteristics in your woods.

Active Management

The second approach to restoring old-growth characteristics involves active management of the forest. Active management can fall along a gradient of intensities that can be tailored to specific goals and forest conditions. Carefully planned forest management provides an opportunity to accelerate the development of old-growth characteristics.

By mimicking the natural processes that have shaped the diversity and conditions of our forests over time, active management can purposefully introduce disturbances that result in old-growth characteristics. Therefore, although it may sound counterintuitive, active management can restore these characteristics and associated values and functions faster than the passive approach.

An active approach can also be used to meet regeneration goals by creating the light conditions necessary for species that are most competitive in partial or even full sunlight (for example, yellow birch, black cherry, white pine, and red oak), restoring aspects of the compositional complexity of old-growth forests. It can benefit wildlife and increase forest resilience by diversifying the number of species present. In addition, active management can provide young forest habitat for those species that depend on it, increase a forest's resilience to such challenges as invasive insects and wind events, and meet important needs for local wood products.

Many of the forest stewardship practices used for meeting your other objectives are also excellent tools for restoring old-growth characteristics; however, it is critical to match these practices with the types of old-growth characteristics you hope to restore.

Restoring Large Trees, Multiple Canopy Layers, Regeneration, and Diverse Species

As with other management approaches, the tactics used for restoring old-growth conditions will vary based on the forest type and existing conditions, given that a guiding principle for ecological forestry is to emulate the natural developmental dynamics for that specific ecosystem. Nevertheless, a key action likely necessary for most of our region's forests is enhancing age-class diversity and spatial complexity. This is because most current forests consist of a spatially uniform arrangement of largely even-aged trees (due to past land use), which is very different from old-growth forests.

Forest management techniques that create gaps in the canopy can be used to encourage the growth of large trees on the edges of the gap, create multiple canopy layers, and establish regeneration. In particular, regeneration harvests can help create the growing space necessary for seedlings to establish. This can be done while retaining a large proportion of the forest in mature trees.

These regeneration methods can be designed to create canopy gaps that emulate those frequently created by small natural disturbances (0.1 to 0.5 acre), although larger openings (1 to 3 acres) with legacy tree retention (see Restoring Large and Old Trees and Deadwood on page 124) can also be created to emulate historic landscape-scale disturbances, like microbursts. Openings of this size can favor shade mid-tolerant (such as oak and pine) and shade-intolerant (such as black cherry, tulip poplar, and bigtooth aspen) species, which are most competitive with higher levels of light. Creating a range of openings in the canopy is critical for common old-growth canopy species such as yellow birch and white pine, which require greater levels of light than shade-tolerant species.

Regenerating diverse species can help you achieve your goals while also increasing forest resilience at the landscape level. This diversity makes forests less vulnerable to challenges such as invasive insects, diseases, drought, and heat, providing multiple ways for the forest to respond to future climate change.

When designing this type of harvest, a forester should allocate 10 to 20 percent of the area in the forest to openings to establish regeneration. This emulates average natural disturbance rates if applied every 10 to 20 years and provides the opportunity to create a mosaic of old forest conditions over time.

Restoring Large and Old Trees and Deadwood

A key element of active management promoting old forest is identifying and retaining trees to live out their natural lifespans in the woods. These trees, often referred to as "legacy" trees, are in the main canopy and are left to serve as future sources of old-growth structure. Unlike trees that are left following traditional timber harvests to grow larger and be harvested in the future, legacy trees are never removed from the woods. Instead, these trees are intentionally left to grow larger and die—providing standing dead trees for habitat—and eventually fall over, providing different habitat as large downed logs on the forest floor. Due to their age, legacy trees can also contain greater levels of genetic variation than younger trees in a second-growth forest owing to the older

ACTIVE PATHWAY TO OLD FORESTS

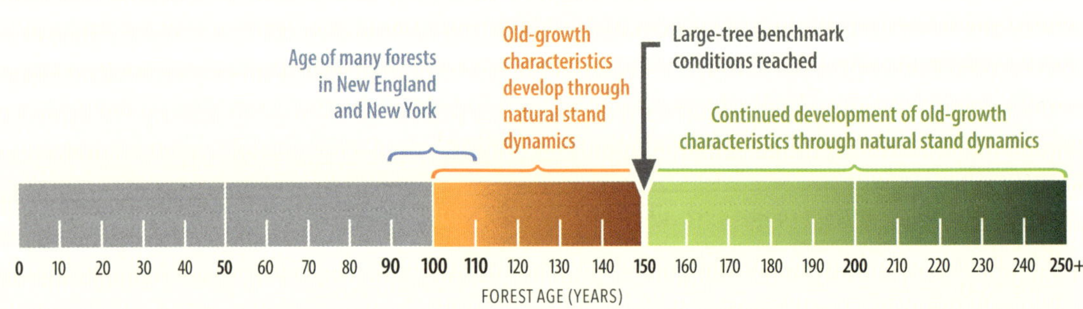

trees' vegetative cells accumulating changes as they age. Such genetic variation is important for our forests as they face an increasing number of challenges.

Legacy tree: a tree purposefully left in the forest to grow large, grow old, and die, creating old-growth characteristics often lacking, such as large living trees and deadwood.

Patch reserve: a group of legacy trees. Also referred to as a "skip," since active forest management practices are skipped in that part of the forest.

Depending on your objectives, individual legacy trees can be dispersed throughout your land or retained in small groups as patch reserves. In the selection of legacy trees and patch reserve locations, preference should be given to trees in your woods that already contain important habitat features, such as cavities and dens, or patches of your woods containing ecologically sensitive features, like a well-developed population of forest herbs or highly decayed deadwood. Likewise, reserving canopy trees with dominant, wide crowns will allow for faster development of large-diameter trees on your property. Finally, selecting long-lived species for legacy trees, such as sugar maple, red and white spruce, yellow birch, red and white oak, white cedar, and red and white pine, can ensure that these old-growth characteristics are present on your land for future generations.

The number of legacy trees left will depend on your landowner goals. If old-growth restoration is your primary objective, then leaving 25 to 50 percent of your canopy trees as legacies and patch reserves will ensure that an old-growth structure will develop over time. If fewer trees are left, it will take longer for the structure to develop. However, leaving even just a few legacy trees per acre can provide old-growth characteristics missing from most woodlots.

The time it takes to restore large living trees can be shortened by removing competing trees directly adjacent to the largest vigorous trees. This increases resources available for their growth and can shorten the time necessary to reach large diameters (more than 20 inches) by several decades. Similarly, large standing dead trees can be created through girdling: cutting rings around the tree through the bark and cambium to stop the flow of water and nutrients, thereby killing the tree. The number and volume of downed logs can be increased by felling trees and leaving them on the ground. These deadwood-creation activities can complement traditional forest management by concentrating on canopy trees not desirable for commercial purposes (often termed "unacceptable growing stock").

Focusing these large-tree management activities between openings and patch reserves can create more varied forest structure and will provide a well-distributed source of large deadwood.

RESTORATION OF OLD-GROWTH CHARACTERISTICS USING ACTIVE MANAGEMENT

Second-Growth Forest (100 years old)

1. There are very few large standing or downed dead trees.
2. The main canopy is uniform without multiple canopy layers. There is little variation in tree size and density.
3. There is very little regeneration.
4. There are few large-diameter trees.
5. Some old-growth characteristics (deadwood, multiple canopy layers, regeneration, increased variation in tree size and density) have begun to develop from natural disturbance.

Second-Growth Forest with Active Management for Old-Growth Characteristics

1. Created a gap of 0.25 to 0.5 acre to diversify the canopy layers, establish regeneration, and increase variation in tree size and density.
2. Intentionally left a large tree on the ground to increase downed deadwood.
3. Thinned trees to increase their growth rates, speeding up development of large-diameter trees.
4. Created a patch reserve in an area with existing old-growth characteristics to serve as a foundation on which to build.

15 Years After Active Management for Old-Growth Characteristics

1. The gap is filled with saplings of diverse species due to the increased light availability, widening the variation in tree size, density, and composition.
2. Downed deadwood has begun to decompose.
3. Residual trees, marked with an "L" (for "legacy tree") to show that they will remain in the forest to live out their natural lifespan, have increased in size due to the thinning.
4. The patch reserve continues to develop.

30 Years After Active Management for Old-Growth Characteristics

1. Young trees continue to grow.
2. Downed deadwood continues to decompose.
3. Residual trees continue to increase in size from the thinning.
4. Trees within the patch reserve continue to develop, contributing to the variation in tree size and density across the stand. The snag falls and now serves as downed deadwood.

SITING OLD-GROWTH RESTORATION AREAS

You may choose to implement old-growth management and restoration, whether passive or active, in all of your woods or only a portion. Choosing a variety of sites with soils ranging from wet to dry will help provide a diversity of old-growth characteristics and conditions.

When considering where on your land to develop old-growth characteristics, it is most effective to identify areas where these characteristics already exist. These areas might include large amounts of downed logs due to a windstorm or a group of large trees containing woodpecker cavities. Areas of special ecological or cultural value (for example, cedar swamps, ridgetops, vernal pools, spring ephemerals, or medicinal plants) should also be considered and may be best suited to a passive approach, providing the added benefit of protecting these important resources.

Establishing a patch reserve around these existing areas will enhance their value as old-growth habitat by providing contiguous forest around them that will be dedicated to developing additional old-growth characteristics. In addition, designating patch reserves in an area with high environmental variation (such as topography, moisture, and soil productivity) will help restore the natural variation in forest types once found across the landscape.

Another consideration in siting your old-growth efforts is directing them to areas of your land in which management for old-growth features will develop fastest. These areas would include those of the highest site quality. Site quality is a way to describe the fertility of an area and is determined by the amount of available water and nutrients for plant growth. High-fertility sites will grow big trees faster than will low-fertility sites. Old-growth structure (big trees, large standing dead trees, and large downed logs) will develop faster on high-fertility sites. Working with a qualified forester will help you understand your options based on the ecological conditions across your land.

While leaving a scattering of legacy trees across your forest will certainly contribute to the restoration of old-growth characteristics, the larger the area dedicated to either the passive or the active approach (or both!), the better. A minimum area of approximately 10 acres in landscapes that are mostly forested is recommended, as this size will provide enough space for the development of environmental conditions and characteristics that are different from the forests around it, which may be stewarded in more traditional ways. However, the minimum size is greatly affected by the landscape surrounding the forest. The more isolated a property is within a non-forest matrix, the larger the area should be.

When to Act

Certainly, there is both regional and global urgency to increase the amount of forests with old-growth characteristics as quickly as possible, given their biodiversity value and potential as natural climate solutions.

However, not every area on your land or in a region is suitable for immediate action. The timing for when you decide to apply old-growth restoration treatments will depend on the condition of your forests as well as the overall timeline for your stewardship.

Older forests that have had time to grow larger trees are more developmentally advanced due to decades of disturbances and already contain some old-growth characteristics. Such forests are excellent candidates for implementing both passive and active strategies, as this will likely serve to speed up the process of restoration. Depending on your ownership, this may apply to all of your forest or only certain areas due to its past land use history. If you own multiple parcels, identifying areas within your ownership that are at this point in forest development will help you find the best locations for the active management approach.

Finally, forest management decisions aren't always triggered by what's going on in the woods. Often they are triggered by other decisions you need to make about your land or by a need or an opportunity related to a community or organization. Therefore, when circumstances have triggered a decision about the way in which your forest is going to be stewarded—whether you are planning a timber harvest or developing a forest management plan—it is a good time to consider the ways in which restoring old-growth characteristics may contribute to your goals.

Since developing these characteristics takes decades, family forest owners should pair these conversations with those about future ownership and use of the land. Formal conservation-based estate planning to ensure that your land remains forested after you are gone provides the opportunity for future old-growth restoration treatments in those areas.

PLACING YOUR GOALS ON A GRADIENT

There is no one specific "old-growth condition" to aim for as an objective and therefore no one way to create it. Instead, it is more valuable to consider increasing the amount of old-growth characteristics in your woods in a way that matches your objectives. While applying the entire set of old-growth restoration practices to your property may be the quickest and most effective way to restore "old growthness" to your forest, this approach may interfere with other management goals, such as creation of early successional habitat and production of sustainable and renewable forest products. It is important to recognize that a gradient of old-growth restoration practices is available to apply to your land. This gradient ranges from doing a single practice, such as retaining a few legacy trees or felling low-quality canopy trees, to employing combinations of practices.

Choosing the level of restoration intensity that is consistent with your other management goals is critical. If you are interested in applying only a single practice, designating legacy trees on your property will provide for the greatest number of old-growth characteristics, such as large old trees, future snags, and logs.

The gradient of restoration treatments can also include a one-time active management approach to "flip the system" followed by a passive approach afterward. Some forests may benefit from this one-time "set it and forget it" active approach to move the forest toward a successional trajectory that will enable natural processes to take over. For example, a plantation of nonnative species could undergo a regeneration harvest to initiate a stand of native species before a passive approach is implemented. Or a stand of diseased beech with a beech understory could be regenerated to diversify the species composition and allow

GRADIENT OF OLD-GROWTH RESTORATION STRATEGIES

Different combinations of practices can be used to restore old-growth structure to your land, and even low levels of restoration practices can be used in areas primarily focused on maximizing timber revenue. Central to all these practices is the use of long-term planning and forest protection.

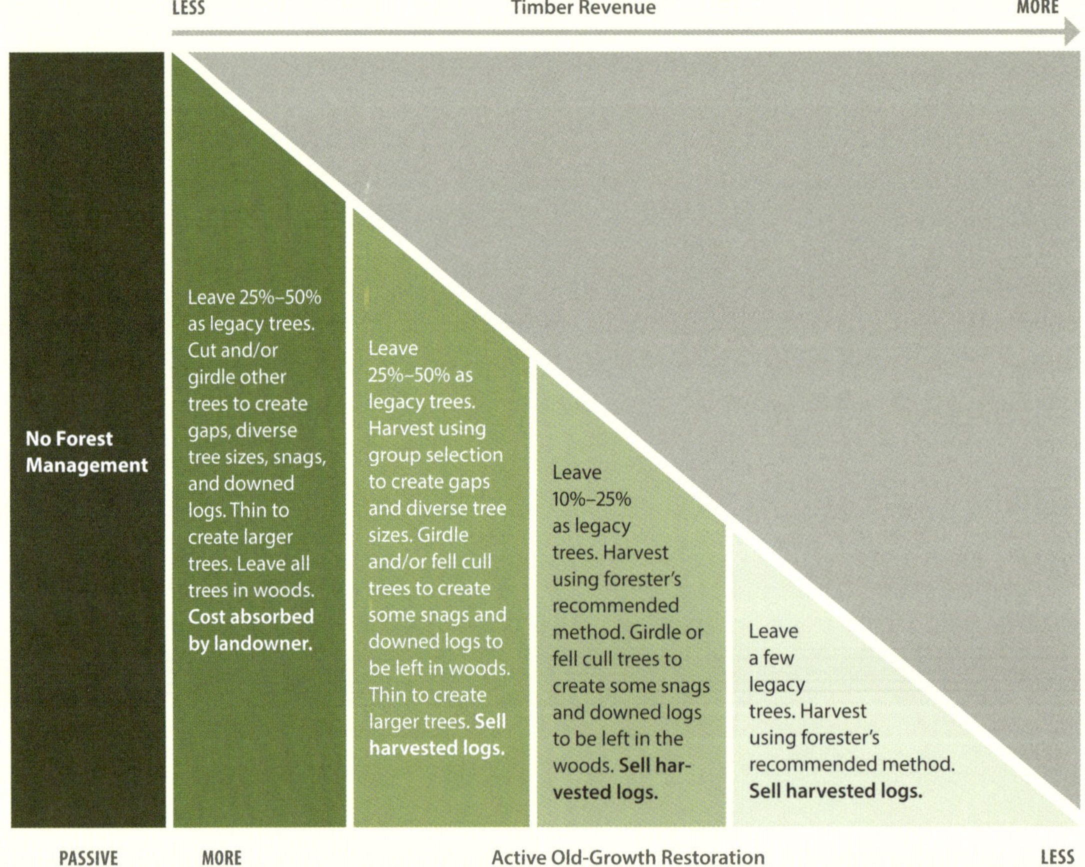

for vigorous trees to establish that will better meet the structural old-growth characteristics.

It is also not necessary to choose one approach for your whole property or, if you own multiple properties, all your properties. A mixed approach can be a great way to meet your overall personal or organizational goals—for example, taking a passive approach in one stand of your property and an active approach in others. At the landscape level, we need both types of strategies to meet our societal goals. Therefore, whichever strategy or combination of strategies you choose will contribute in meaningful ways.

YOUR LAND IN CONTEXT

Another important consideration in planning for old-growth restoration is how your land fits into the surrounding landscape. Like most management objectives, an excellent way to increase the functioning of the areas containing old-growth restoration treatments is to implement these strategies in landscapes where other landowners, both public and private, are doing similar management. Even if multiple landowners in a landscape each implemented just a small amount of old-growth restoration strategies, it would add up by providing stepping stones across the landscape for animal and plant species to move.

Consider coordinating your management activities with adjacent landowners to increase the size and effectiveness of these areas serving to replicate old-growth conditions on the landscape. If the property lies within a forest matrix, a smaller area dedicated to these strategies can contribute to the landscape, but if the land around your forest has been converted to residential development, then dedicating a larger area to these strategies is recommended.

Though it is natural to focus attention on property-level considerations, since that's the scale on which most landowners are making decisions, taking a landscape perspective is not only important in regard to restoring old-growth characteristics but also critical to the overall resilience of our forests. Natural disturbances don't start and stop at property boundaries. Landscapes with a diversity of species composition and forest structure will help ensure that our forests can adapt to challenges such as climate change and invasive insects.

Ultimately, we need both passive and active approaches on the landscape. The passive approach provides ecological benefits that the active approach cannot, such as maximizing carbon storage. Likewise, it's important to acknowledge the value of areas where forests are managed with a greater emphasis on young forest habitat for species in decline or for wood production to meet local wood and other forest product needs instead of relying on other regions or parts of the world. These approaches complement each other across the landscape. Whichever approach meets your needs, keeping your forest as forest provides immense public value and supports myriad plants and animals, especially in the face of global change.

CASE STUDY

Restoring Old-Growth Characteristics Within a Connected Network of Lands

The Nature Conservancy Appalachians Program

LOCATION: Resilient and Connected Network of Lands in the Appalachians (RCN)

Our shared land use history of clearing forests for agriculture means that ecologically young forests exist across our entire region. That means that restoring old-growth characteristics to our forests will necessitate individual landowners of all types implementing restoration strategies across a very large landscape. How does that happen when there are so many different landowners?

Extending from northern Alabama through the Canadian Maritimes, the Appalachian region forms a connected corridor of forest across a wide gamut of climates and biodiversity. A team of The Nature Conservancy (TNC) conservation practitioners from across the Appalachian landscape, in partnership with the Appalachian Mountains Joint Venture (AMJV), are working on an initiative to accelerate the restoration of old-growth characteristics across the region within a variety of forest types. The goal of this management is to restore elements that are currently missing from our ecologically young forests. Forest manager interest in late successional habitat management is high, as is political will, but there is little formal guidance, funding, or community of practice to learn from. To advance this style of management across the landscape at a meaningful scale, TNC hopes to foster resilience and connectivity using forests themselves as the template and building a coalition of conservationists, foresters, and land managers to address these needs across the region.

These efforts include not only developing on-the-ground forest management strategies to restore old-growth characteristics on land owned by TNC but also collaborating with conservation partners across the landscape, such as Pennsylvania Game Commission, Maryland Department of Natural Resources, and the Ruffed Grouse Society. Several projects are already designed and are in different stages of implementation in Pennsylvania, Maryland, West Virginia, Tennessee, and Kentucky. At the property level, these strategies focus on enhancing growth conditions for desired canopy trees, increasing the structural complexity of the forest, and regenerating species that will be well adapted to future climate conditions. These strategies vary by forest type across the region, as different forest types have different tree species and different historic disturbance patterns. At a landscape scale, the conservation

strategy for the Appalachian region includes increasing the amount of old-growth characteristics across this multistate region. Doing so will help increase forest resilience and biodiversity and weave the parcellated landscape back together through a critical mass of old-growth characteristics across properties.

Success at a scale as large as the Appalachian region is no small task. Their implementation strategy starts with the establishment of demonstration sites to provide proof of novel management concepts on TNC and partner-owned properties. The demonstration sites will help refine strategies to restore old-growth characteristics in different forest types and provide tangible on-the-ground examples of successful management implementation. Once established, these demonstration sites will be used as anchor points within the landscape from which to reach out to conservation partners with overlapping interests, such as the family forest owners who own and control the majority of forest in the region. TNC and their conservation partners are hoping to bring science-based strategies and experience, as well as outreach capacity, to reach family forest owners and conserve the Appalachian region one property at a time.

As a landowner, your land is a critical part of large landscapes that are essential for maintaining diverse and resilient forests. Working with partners in the landscape is a highly effective way to amplify the impact of forest stewardship, as each owner's efforts leverage the others', leading to a whole greater than the sum of its parts.

Working with partners in the landscape leads to a whole greater than the sum of its parts.

Nature Conservancy foresters meet in the field with foresters from the Tennessee Wildlife Resources Agency to plan forest stewardship that will encourage the development of old-growth characteristics.

LONG-TERM PLANNING

Developing old-growth characteristics takes a commitment to forest continuity and stewardship—that is, allowing your land to remain forested and intentionally leaving more trees in the forest. Even with active management, old-growth characteristics take decades or longer to develop, requiring the need for long-term planning, both within the forest and, in the case of family forest owners, out of the forest through conservation-based estate planning.

Within the Forest

Within-forest planning is necessary to document your stewardship approach and intention over the long term. Legacy trees and patch reserves should be clearly designated (for example, marking legacy trees with an "L" and patch reserve boundaries with paint or scribing) and described both on paper (a map, forest management plan, or baseline documentation report) and in digital form (photos and GPS locations of legacy trees and areas being managed for old growth). Long-term forest management planning can ensure that future decisions about your woods will not disrupt the development of old-growth structure in areas that have been dedicated to restoring old-growth characteristics.

Conservation-Based Estate Planning

It is very important to recognize that the development of old-growth structure will depend on what happens to your land once it is passed on. If your forest won't be there in one hundred years, what you do now won't matter. Therefore, talking to your heirs and using conservation-based estate planning tools such as conservation easements are excellent ways to ensure that your woods will stay forested long enough to develop the old-growth characteristics you desire. This topic will be covered in Chapter 12.

GETTING STARTED

Begin by thinking about how developing old-growth characteristics fits with your other goals for your land. Working with a forester to evaluate your land, its landscape context, and your options for developing old-growth characteristics is an excellent way to identify the appropriate management options and move forward. Contacting your local land trust or public conservation agency to find out your land conservation options is another important step.

Public and private conservation organizations interested in restoring old-growth characteristics can start by determining where the organization falls on the old-growth restoration gradient. This will help you determine target percentages for restoration. Identifying and documenting areas of your forest with existing old-growth characteristics and those with special ecological and cultural value will help you identify specific areas in your forest to focus on and build off of to reach your overall target percentages. Decide the approach(es)—active, passive, or both—that best meet your organization's goals and fit the conditions of your forest and its landscape context. If using an active approach, choose regeneration harvests that will meet your structural and species composition goals.

FIELDWORK

What Are the Possibilities for Restoring Old Growth on Your Land?

Determining your level of interest in restoring old-growth characteristics will help you decide how much area to designate for retention as legacy trees and patch reserves. Is restoring old-growth characteristics a primary, complementary, or secondary goal for your forest? How does this goal fit with your other goals?

Evaluate the current "old growthness" of your forest. What types of structures do you see that are characteristic of an old forest?

Evaluate your landscape context. How much old growth do you think there is in the land surrounding yours? What is its spatial relationship to your forest?

Evaluate your current plan for the future ownership and use of your land. Have you formalized your wishes in a way that will ensure your forest has the time to develop old-growth characteristics?

Practicing Ecological Forestry to Restore Old-Growth Characteristics

CONTINUITY: Identify current old-growth characteristics in your forest to be retained through future forest management. Retain additional biological legacies including legacy trees, patch reserves, and deadwood throughout the forest. Engage in conservation-based estate planning to ensure your forest will be retained as forest long enough to develop old-growth characteristics.

COMPLEXITY AND DIVERSITY: Increase vertical diversity by promoting multiple layers of trees and vegetation; increase horizontal diversity by encouraging patchiness within the forest—gaps, patch reserves, and thinned trees. Plan for future inputs of both standing (snags) and downed (logs) deadwood. Increased structural complexity will lead to increases in biodiversity as the forest becomes a more niche-rich environment, including supporting fungal communities through increases in deadwood abundance. Strive to meet regeneration goals for species well adapted to the site and predicted to do well in the future.

TIMING: Some characteristics take time to develop. Extend the timing between regeneration harvests to give more time for trees to grow and age. Give patch reserves time to establish and spread understory plant communities. The time it takes to develop other characteristics, such as large trees, can be speeded up through active management.

CONTEXT: Consider prioritizing restoration in areas of your forest that will connect or be close to other properties with old-growth characteristics. If possible, prioritize the development of old-growth characteristics in stands on your property that may not be represented within the landscape (for example, unique forest types or stand conditions).

> "What you do makes a difference, and you have to decide what kind of difference you want to make."
>
> —Jane Goodall

CHAPTER 7

Mitigating Climate Change

CLIMATE CHANGE CAN SEEM LIKE AN OVERWHELMING CHALLENGE. The good news is that as a forest landowner you can make a significant difference. The land use choices you make—specifically your decisions about the future use and management of your forest—play an important role in reducing the effects of climate change, regionally and globally.

Forests are an essential natural solution for climate change. They remove carbon dioxide from the atmosphere and store it in various ways. The decisions about carbon and the management of your forest come with trade-offs. Making an informed decision requires a good understanding of forest carbon and the influence you can have on it.

Soils are typically the largest pool of stored carbon within the forest.

FOREST CARBON BASICS

Forests take in carbon dioxide from the atmosphere to make energy through photosynthesis. Trees then use this energy to thrive and grow. Through this process, trees capture carbon in the form of wood and other organic matter, like bark and leaves. In fact, one half of a tree's weight consists of stored carbon. Since most of our region is forested, our landscapes play a globally important role in both sequestering and storing carbon and ultimately helping to reduce the impact of climate change.

Most of our region's forests are owned by families and individuals. Therefore, the decisions landowners make about their forest will have the greatest impact on the amount of carbon our forests absorb and store. There is also a significant amount of our region's forest land that is owned by public agencies, conservation organizations, tribes, and corporations; all of their actions can also reduce the effects of climate change. Because every forest management strategy comes with trade-offs, meeting all of society's needs while mitigating climate change will require a mix of passive and active strategies across the region. What role will your forest play?

BASIC CARBON TERMINOLOGY

A carbon pool is a part of the forest that stores carbon and can accumulate or lose carbon over time. Carbon pools include live aboveground biomass, such as trees, soil, and organic matter. There are two basic aspects to a carbon pool: how much it contains (called carbon storage), and how much it is changing (called carbon sequestration).

The terms *storage* and *sequestration* are often used interchangeably; however, each one has a specific meaning and reaches its maximum level at different times in forest development. Nevertheless, both are necessary for reducing the effects of climate change.

Carbon storage is the amount of carbon that is retained in a carbon pool within the forest. Storage levels increase with forest age and typically peak when forests in the northeastern United States are more than two hundred years old.

Carbon sequestration is the process of removing carbon from the atmosphere for use in photosynthesis, resulting in the maintenance and growth of plants and trees. The rate at which a forest sequesters carbon changes over time. In our region, carbon sequestration typically peaks when forests are young to intermediate in age (around 30 to 70 years old), but they continue to sequester carbon throughout their entire lifespan.

WHERE IS CARBON STORED IN A FOREST?

A forest stores carbon in different pools, and the amount of carbon in pools changes over time. Factors that influence the amount and proportion of carbon in a given pool include the age of the forest, the species of trees making up the forest, natural and human disturbances, soil characteristics (such as texture and drainage), and the land use history of the forest.

In addition to changing over time within and among the pools, carbon storage varies among forest types. This variation is strongly influenced by the climate in which the forests grow. Generally, the warmer the climate, the longer the growing season and the greater amount of carbon stored aboveground in trees. Relatively less carbon is stored in the soil in warmer climates, due to faster decomposition rates. The table on page 142 shows the average amount of carbon stored in different pools of our current 80- to 100-year-old forests, in forest types common to our region.

CARBON STORED IN 90-YEAR-OLD FORESTS BY FOREST TYPE (TONS OF CARBON/ACRE)

FOREST TYPE	LIVE TREE	DEAD WOOD	FOREST FLOOR	SOIL ORGANIC	TOTAL (TONS/ AC)
Maple-beech-birch (NE)	46 (47%)	16 (16%)	5 (5%)	31 (32%)	98
Spruce-fir (NE)	23 (31%)	5 (7%)	2 (3%)	44 (59%)	74
Oak-hickory (NE)	47 (51%)	17 (18%)	4 (4%)	24 (26%)	92
Aspen-birch (NLS)	29 (28%)	6 (6%)	5 (5%)	65 (62%)	105
White-red-jack pine (NE)	36 (42%)	9 (10%)	6 (7%)	35 (41%)	86

NE = Northeast
NLS = Northern Lake States

THE TREE VERSUS THE FOREST

When trying to understand the amount of carbon that might be sequestered and stored by a given forest, a common point of confusion lies in the assumption that the growth of a forest follows the same pattern as the growth of an individual tree. In fact, the two are quite different.

Individual trees growing within the main canopy of a forest gain biomass at an increasing rate as they age. Their growth continues until they reach old age (greater than two hundred years old), at which time their growth rate slows, leading to the natural decline and death of the trees. This is a pattern long recognized by forest scientists and recently popularized, as the important role that carbon storage and sequestration play in mitigating climate change has become a more prominent forest management objective.

However, growing space and resources are finite in a forest, so not all trees within a particular forest can grow at an optimal rate over time. While vigorous individuals grow in size and dominance, less vigorous trees have slow growth rates due to a lack of space and resources. So even though some trees (those in upper-canopy positions) continue to grow at high rates until old age, many do not. The result is a reduction in the growth and sequestration rates of the forest as a whole.

The ability of these dominant individuals to continue growing is an important attribute to consider when objectives include restoring large-tree habitat or developing large-diameter sawlogs, but it should not be confused with forest-level growth and carbon sequestration rates, which generally decline with age, regardless of the tree species or soil conditions. Despite lower forest-level sequestration rates as the years go by, the forest continues to increase its level of carbon storage.

HOW FOREST AGE AFFECTS CARBON

A forest goes through stages of development on its way from a seedling forest to a late successional forest (see the successional clock on page 145). As forests grow older, the species within them shift from those that need full sunlight to grow (shade intolerant) to those that compete best in partial sunlight (shade mid-tolerant) to those that are most competitive in full shade (shade tolerant). Each stage of forest succession and development provides unique benefits based on the forest's structure and composition. For example, young forests provide one type of wildlife habitat, and old forests with large-diameter trees provide habitat for a different suite of species.

Similarly, a forest's maximum rate of carbon sequestration happens at one stage of forest development, when tree diameter ranges from approximately 4 inches to 16 inches. The maximum amount of carbon storage happens at another stage—when trees are greater than 20 inches in diameter. The age of the forest strongly influences both the rate at which forests sequester carbon and the amount of carbon that they store.

Young Forests: Maximizing Carbon Sequestration

As discussed in Chapter 1, after a disturbance forests typically regenerate themselves naturally through seeds and sprouts. Seedlings and sprouts grow into saplings (less than 5 inches in diameter), and as saplings grow there is tremendous competition for resources. The saplings grow vigorously until their crowns grow into one another and occupy all available growing space. Trees that can grow faster than neighboring trees become dominant and claim more and more space to grow and survive. Those trees that lose space are outcompeted and eventually die. The space and resources vacated by these dead trees are increasingly used by the remaining trees. In this way, the resources of the site continue to be concentrated into fewer and fewer trees that grow larger and larger. This is the stage of forest development at which the rate of carbon sequestration is highest, as the amount of leaf area and the rate of photosynthesis peak during this period of high tree-to-tree competition.

These higher rates generally occur when the forest is approximately 30 to 70 years old or the trees are approximately 4 to 16 inches in diameter, though the specific age and size will depend on such factors as site quality and land use history. Soon after the forest canopy closes, the overall growth of the forest slows down and, with it, the sequestration rate. However, trees continue to sequester significant amounts of carbon to continue to grow and maintain themselves.

One important thing to recognize is that the forest might actually be a source of carbon immediately following a disturbance, as rates of tree growth, although rapid, are unable to counteract losses of carbon due to the decomposition of organic matter in the soil. This loss of carbon from decomposition is enhanced when large openings are created in the forests, which increase soil temperature and moisture availability and hence microbial activity. It generally takes 10 to 15 years before there is enough forest growth to shift a disturbed area from a carbon source to a carbon sink.

Old Forests: Maximizing Carbon Storage

As a forest ages, the total amount of its stored carbon increases as carbon accumulates in the different pools. Trees grow larger in height and diameter, increasing the size of live aboveground pools. As trees get larger over time, their roots grow and spread, increasing the size of soil organic pools. At the same time, larger trees drop more leaves and branches on the ground, adding to the litter pool. As the forest ages, trees die due to insects, disease, wind and ice storms, and competition. As they die, the deadwood pool grows in the form of snags and downed logs. The litter and dead trees all contribute to the soil pool over time. Together, these processes increase an older forest's ability to store carbon in the various pools. This is the stage of forest development with the highest amount of carbon storage.

Old-growth forests can provide us with a guide as to how much carbon mature forests store. Estimates of the carbon stored in these forests range from 100 to 120 metric tons of carbon per acre. Due to our land use history, our current forests are relatively young, many around one hundred years old, and generally store 60 to 80 metric tons of carbon per acre. Carbon in our current forests accumulates at a rate of about 0.41 metric tons per acre each year in a typical maple–beech–yellow birch forest. Given this rate of carbon accumulation, our current maple–beech–yellow birch forests will need to continue growing at this rate, without a major forest disturbance, *for about another one hundred years* before they would have the level of carbon storage that we find in old-growth forests. Future gains in forest carbon will primarily come from the diameter growth of trees, additions to the deadwood pool from dying trees, and the accumulation of soil organic carbon from root growth and decomposition.

FOREST SUCCESSION AND CARBON

Let's look at the successional clock again to see how a forest's rate of carbon storage and sequestration are closely linked to the age of the forest. The numbers on the successional clock represent the age of the forest in years.

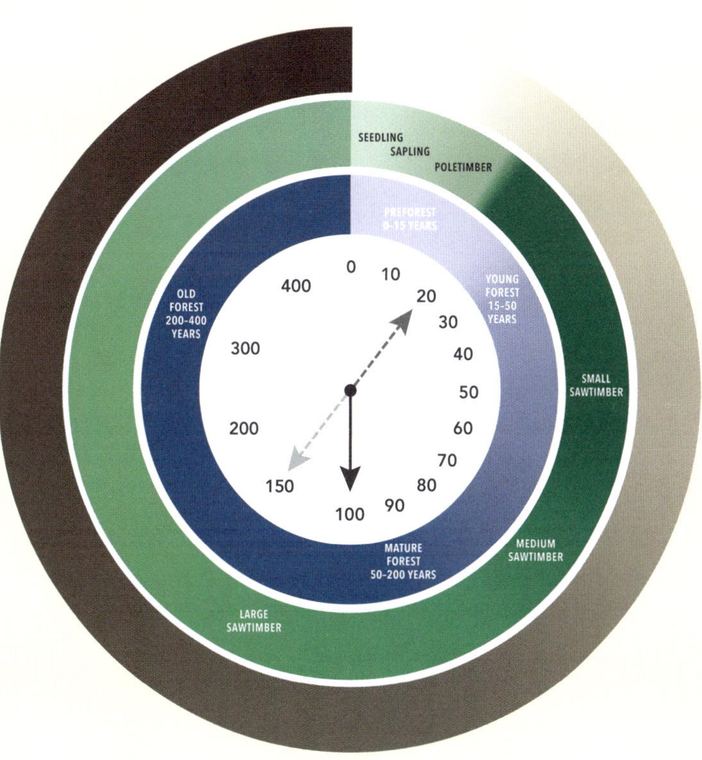

LEGEND

- **Changes in carbon storage over time.** The darker the brown, the more carbon storage.
- **Changes in carbon sequestration over time.** The darker the green, the more forest-level carbon sequestration.
- **Forest stages over time.**

Multi-Aged Forests: Balancing Carbon Sequestration and Storage

It's easiest to describe all the trees in a forest as being the same age—that is, all young or all old. In fact, many of our forests are the same age (even-aged), due to our land use history of agricultural clearing and forest harvesting. However, the older our forests get, the more they will shift from even-aged forests to multi-aged forests. In our region, it is common for trees to die individually or in small groups due to frequent low-severity disturbances such as wind, ice, and insects, or to partial harvests, which do not remove all the trees at once. These types of partial disturbances leave some older trees while creating the opportunity for seedlings to establish themselves and saplings to grow in the gaps created by the death of canopy trees. Some forest management strategies, such as single-tree and group-selection systems, mimic the

ISN'T CARBON STORAGE THE MOST IMPORTANT FOREST BENEFIT?

While forests serve a critical role in the mitigation of climate change, we must balance this benefit in a way that sustains all of the benefits forests provide. The Spanish-American philosopher George Santayana famously wrote, "Those who cannot remember the past are condemned to repeat it." There was a time when forests were valued only for their ability to produce wood products, leading to strategies that simplified forest structure and species composition and reduced the benefits they provide. In fact, to this day, forestry continues to live with the stigma of its timber-focused past. Now, there is growing pressure to reduce our management of forests to only that which will maximize their ability to mitigate climate change, yet this comes with the risk of repeating the same mistake we made more than a century ago with timber, but now with a singular carbon focus.

Rather than focusing on one benefit, no matter its importance, ecological forestry strives to restore fully functioning forests and, with them, the full suite of benefits. Within this context, compared to standard forestry, the principles of ecological forestry provide for increased and long-term carbon benefit to mitigate climate change.

region's frequent low-severity disturbances and thus create multi-aged forests.

Forests composed of trees of various ages have the combination of characteristics that the trees of each age possess. For example, a forest with equal areas of young trees and old trees will have high rates of sequestration from the younger trees while maintaining the storage capacity and sequestration rates of the surviving older trees. When making decisions about your forest management strategy, one consideration is whether you want to maximize carbon storage, sequestration, or a combination of the two.

LAND USE DECISIONS AND CARBON

With a greater understanding of forest carbon, it's now possible to better consider the implications of land use decisions on forest carbon sequestration and storage. Two of the most influential decisions that landowners make regarding forest carbon are (1) whether their forest will be converted to non-forest use by them or a future owner, and (2) whether to engage in active forest management—and, if so, how to do it in a way that meets their goals.

Land Use Decision 1: The Future of the Land

The first and most significant decision landowners make surrounds what will happen to their forest in the future. Will it be converted to another use, such as residential, commercial, agriculture, or energy development (solar fields, pipelines), by them or a future owner, or will they decide to keep their forest as forest and make the necessary plans to ensure this?

Forest Conversion to Non-Forest Land Uses

Conversion of forests to non-forest land use affects all of a forest's carbon pools. It eliminates most of the carbon storage and all of the forest's capacity to store and sequester carbon in the future. Forest conversion eliminates the live aboveground, deadwood, and litter pools of the forest by removing them from the site. Soil disturbance from stumping, grading, and plowing decreases the soil carbon. In addition to these losses, conversion of forests to other land uses is often permanent, or at least lasts decades, meaning that the carbon that was lost from these forests is generally not recovered. In addition, forest conversion means not only a loss of carbon sequestration and storage but also a loss of all forest benefits (habitat, clean water, local wood products, etc.). Though some forest loss may be necessary to achieve personal and societal goals, remember that the most significant losses in forest carbon are the result of conversion of forest to non-forest uses.

Keeping Forests as Forests

Keeping forests as forests is the single most important action a landowner can take to maintain forest carbon sequestration and storage. It is therefore the single most important action a forester or natural resource professional can facilitate. There are several conservation-based estate-planning tools that can be used to help family forest owners keep their land in its forested condition, which will be covered in Chapter 12.

Afforestation

If the biggest loss of carbon sequestration and storage comes from conversion of forests to other land uses, then the biggest gain is the reversion of non-forest land back to forests. If forest carbon is your primary goal, then one option is to allow non-forest land uses, such as abandoned or unproductive fields, to revert back to forest. However, the balance between maximizing forest carbon storage and sequestration and accommodating the need for local agriculture is an important trade-off to consider.

Land Use Decision 2: Forest Management

A second major decision you will make that impacts forest carbon is the forest management strategy you choose. Both the passive and active approaches have implications for carbon, and both have trade-offs.

Passive Forest Management

Many landowners choose to adopt a passive approach to their land by not engaging in timber harvesting and letting nature take its course. Though a passive approach to forest management means no timber harvesting, it can still be active in terms of other activities that can help increase forest resilience, such as invasive plant control. Assuming a forest is able to grow and develop without significant challenges (invasive plants, significant deer browse), the passive approach will likely maximize forest carbon storage through the accumulation of carbon in each pool as the forest grows older, but it will not maximize the carbon sequestration rate, which, as previously described, occurs in younger forests.

Having forests within our landscapes that are allowed to accumulate and store high amounts of carbon is a critical part of reducing the impact of climate change. Since sites with ample water and nutrients will grow larger trees than those with less water and fewer nutrients, forests on these richer, more fertile sites will be able to grow and store more carbon. These fertile sites are also often the best growing sites for active forest management, so you will need to consider how to prioritize your goals or find areas of your forest for both passive and active approaches. In addition to site quality, large areas of forest with low fragmentation may be more resilient to disturbances and therefore better able to store carbon than smaller, fragmented areas.

Considering the Carbon Trade-Offs of the Passive Approach

Forests provide many essential benefits, including carbon, but not always in equal proportion. Choosing a strategy for your forest may mean that some benefits are enhanced while others are reduced. These are decisions that every landowner must make, hopefully after fully understanding the trade-offs. Climate change is a critically important issue. Taking a passive approach to forest management will likely provide the greatest amount of carbon storage. However, there are other important trade-offs of the passive approach to consider.

Forest resilience is one potential trade-off of passive management. Forest conversion and timber harvesting are not the only ways in which forests lose carbon. One of the anticipated impacts of climate change is more frequent and more severe natural disturbances, such as wind, ice storms, and wildfire. In addition, invasive species and deer overpopulation pose an increasing threat to our forests. Typically, these forest disturbances disproportionately affect one part of a forest (wind events affect the trees with the biggest crowns; insects affect certain species of trees). Active management can make our forests more resilient to these disturbances by increasing species and structural diversity, thereby reducing the risk that a disturbance will kill all the trees in a forest. This includes restoring more complex forest structures and reducing fuel loads in fire-adapted forests to reduce wildfire risk. Forests with diverse conditions contain multiple mechanisms for recovery following such events, which will allow for carbon to return to pre-disturbance levels more quickly.

Though active forest management would temporarily reduce the amount of carbon stored in the forest, offset somewhat by wood products, it may help prevent an even larger reduction in carbon storage by avoiding losses due to a large-scale disturbance. Forests in landscape positions vulnerable to natural disturbances (such as exposed or drought-prone sites), even-aged forests, forests dominated by a limited number of species, and forests with a high proportion of species with known forest-health issues (such as ash and hemlock) or high fuel loading are most susceptible. We will talk more about forest resilience in Chapter 9.

Maintaining populations of native wildlife is a common goal of many landowners, both public and private. Approximately 80 percent of our region's vertebrate wildlife species rely on forests of different ages for different parts of their life cycle. Therefore, to maintain native populations there needs to be enough forest habitat of different ages to support these populations.

Because most forests in our region are around one hundred years old, species that rely on the currently limited amount of young forest (such as the chestnut-sided and golden-winged warblers and the New England cottontail) are in significant decline. Creating young forest habitats for these declining species through active forest management means sacrificing carbon storage at the property level, but it will have a disproportionate positive impact on regional biodiversity. Similarly, active management strategies for increasing old-growth characteristics such as large trees will reduce overall carbon storage in the near term but will accelerate the development of this currently rare forest stage. Ultimately, to sustain our native wildlife populations, we need a range of forest ages and types across the landscape.

The chestnut-sided warbler is a disturbance-adapted species that depends on young forests.

CARBON CREDITS:
A Future Carbon Option?

Selling the carbon your forest sequesters and stores typically requires implementing stewardship practices that retain carbon within the forest, such as by preserving a certain number of trees.

Your forest provides many critical functions for society, including removing carbon dioxide from the atmosphere through carbon sequestration and storing it, thus reducing the impact of climate change. There are ongoing efforts to establish programs to pay forest landowners for providing this service. To sell carbon credits (one credit is equivalent to one metric ton of carbon dioxide), a landowner typically needs to meet several requirements: verifying that the forest is sustainably managed, providing a detailed inventory of the amount of carbon in the forest and future projections of growth, and signing a long-term contract to ensure that the forest remains a forest and will not be harvested in a manner that reduces the amount of carbon stored in it. Meeting these requirements and earning a profit from carbon credits typically requires large amounts of acreage. This acreage can consist of one large property or a number of smaller ownerships.

Sustainable wood products play a role in mitigating climate change as well. Maximizing forest carbon storage and sequestration is only part of the global carbon picture. To understand the full role of forests in the global carbon cycle, it is critical to consider both the amount of carbon stored in forest products and the amount of carbon that is saved when wood is used in place of more carbon-intensive materials, such as steel and concrete.

All carbon removed from the forest during a timber harvest is not immediately returned to the atmosphere. Approximately one third of the forest products harvested in the northeastern United States are made into products such as furniture, flooring, and dimensional lumber (such as 2×4s) with long lifespans. Forest management provides the opportunity to improve the quality of forest products by concentrating growth on higher-quality trees, which will increase the number of trees used for the wood products that will store carbon for long periods of time.

If we choose not to use wood, what are the carbon costs of substitute materials? How much energy does it take to acquire these materials? How much energy does it take to convert them into a usable product? The graph below compares the energy costs of common building materials across their entire life cycle, from harvest to disposal. The numbers help highlight the important role wood plays as a renewable, environmentally friendly building material, with climate and carbon advantages.

If we decide to continue using wood because it is environmentally friendly but do not harvest it locally, it must then come from somewhere else. If it comes from outside our region, it takes energy and produces carbon emissions to bring those wood products to our region. In addition, the places from which we typically import wood may have the potential to store more carbon than our region (Pacific Northwest) or may have less environmental oversight protecting forest resources (other countries).

MANUFACTURING ENERGY OF COMMON BUILDING MATERIALS, NORTHEAST AND MIDWEST US

The energy cost of wood products is just over half that of cement and a small fraction of the energy cost of metals, highlighting the importance of wood as a renewable, energy-efficient resource.

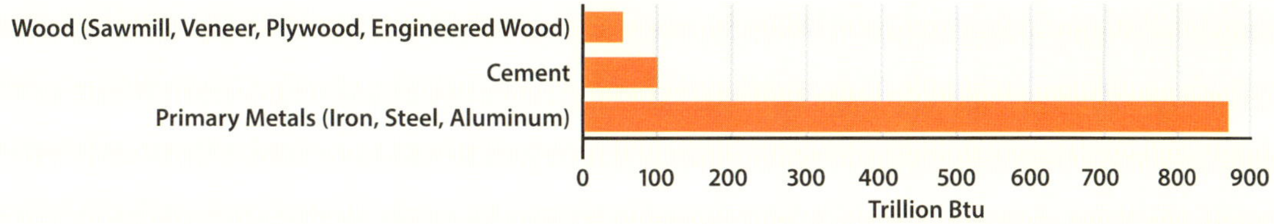

US Energy Information Administration, 2018 Manufacturing Energy Consumption Survey

Understanding the whole forest carbon story necessitates looking beyond the property level to both the regional and global scale and includes considering the many roles forest products play.

Active Forest Management

Many landowners choose to implement some form of active forest management, including timber harvesting, to achieve their ownership goals. Active forest management can also generate income, which may help pay taxes and other costs necessary to keep the land forested. The following are descriptions of the implications of active forest management on the two biggest carbon pools in a forest: live aboveground and soil. There are also recommendations for deadwood pools, which hold the potential for significant increased storage of carbon in the future.

- **Live aboveground carbon:** All harvesting reduces carbon storage of a forest below the maximum potential for the site. However, timber harvesting also often results in the establishment of a new cohort of young trees that can increase the carbon sequestration rate of the forest. The amount and type of trees removed, as well as the timing, will determine the specific impact on carbon.
- **Soil carbon:** Harvesting has little appreciable impact on soil carbon as long as soil disturbance is minimized and all the slash (tree debris left after harvesting), treetops, and nonmerchantable lengths are not removed. Best management practices (BMPs) should be used to protect soil during forestry operations.
- **Deadwood:** Since dead standing trees and downed logs have little to no merchantable value, they are typically not removed from the forest during a timber harvest. Timber harvests can be implemented to maximize the amount of deadwood left on-site, such as slash, treetops, and nonmerchantable lengths of logs.

Considering the Carbon Trade-Offs of the Active Approach

The most important carbon consideration of active forest management is the loss of forest carbon storage resulting from the removal of trees. Though some of the trees removed during a timber harvest will end up in long-term forest products, any removal of trees is a temporary reduction in carbon storage on that property at that time. This reduction can be minimized by applying strategies that reduce carbon loss.

The effects of active forest management on carbon storage are often considered only at the property level. However, it is also important to consider the effects at the regional scale. Our region has more forest growth in a given year than is removed for wood products. In other words, for every unit (measured in cubic feet) of wood that is removed from the region's forests as part of timber harvesting, a greater amount of wood (and therefore carbon) is accumulated in each state. This means that while forest management reduces carbon storage at the property level, from both a statewide and a regional perspective, our region continues to grow more wood (and carbon) than it harvests. The amount of wood removed and grown, and therefore the amount of carbon sequestered and stored, changes over time as landowners make decisions about the management of their forests.

CARBON-INFORMED FOREST MANAGEMENT

If you choose to move forward with some type of active forest management on your land, there are strategies you can implement to reduce the loss of carbon storage from the forest while increasing carbon sequestration and resilience. Since the soil and live aboveground carbon are the largest pools within the forest, and the deadwood pool has significant potential to increase, focusing on strategies for these pools will have the greatest impact.

Soil Pool

If you choose to do a timber harvest, best management practices (BMPs) will help protect your forest's soil and water health by avoiding soil damage, such as rutting and compaction, and controlling the overland flow of water, which can cause erosion. See Resources (page 264) for more information about your state's forestry BMPs. A consulting forester will help you do the following:

- Develop a strong contract with the logger that specifies soil and water performance standards for the timber harvest (such as rutting depths and timing of harvest).
- Lay out logging roads to minimize their number and ensure that they are located on stable ground.
- Avoid soil disturbance by timing the harvest to frozen or stable conditions.
- Monitor the harvest to make sure that the ground is stable enough for harvest operations and that the contract provisions are being followed.
- Conduct a final inspection before the timber harvesters leave the site to make sure that the site has been stabilized and the contract has been satisfied.

Live Aboveground Pool

Carbon-informed forest management must consider both the structure (size, number, and arrangement of trees) and the composition (species of trees) of a forest. Also keep in mind that both carbon sequestration and storage are important for maximizing a forest's role in mitigating climate change. Active management offers the opportunity to establish a desirable balance of small, young, fast-growing trees for high sequestration rates and large, old, slow-growing trees to maintain carbon storage. Ecological forestry's focus on continuity, complexity, and timing will increase the carbon benefits of your forest.

Large trees store a disproportionate amount of aboveground carbon and are also important habitat features within a forest.

Maintain Large Trees

Large-diameter trees make up a disproportionate amount of the live aboveground carbon stored in a forest. In addition to their carbon-storage benefits, these trees are excellent for providing wildlife with cavities and food, they have high aesthetic value, and they may be an important seed source for future trees. There are several ways to grow and maintain them while practicing ecological forestry:

- Designate large trees to leave permanently standing. They will remain in your forest's live aboveground carbon pool and eventually become part of the deadwood pool. These "legacy trees" can be scattered as individuals across the forest or left in small groups of at least a quarter acre in size. Groups of legacy trees can be placed around areas of high ecological value, such as vernal pools or other sensitive sites, to support wildlife.
- Maximize a tree's ability to store carbon by letting it grow larger before cutting. For planned timber harvests, grow vigorous trees an extra 10 to 20 years past your harvest timeline or 2 to 3 inches larger than your target diameter. Sometimes harvests are unplanned, triggered by events that do not allow the timber harvest to be delayed. In these cases, consider leaving additional retention trees on-site.
- Ask your forester about using regeneration methods that maintain large trees across the forest.

Ensure Regeneration

Establish a new age class of trees by making sure your forest has a chance to regenerate. This may mean looking for ways to address invasive plants and excessive browsing by deer and moose. Timely regeneration of species well suited to the site and future conditions will ensure that there are trees in place to sequester and store carbon into the future.

Diversify Tree Ages

Identify the appropriate combination of young and old trees to meet your goals, and develop forest resilience through diversity.

As previously described, carbon sequestration rates peak when forests are young and then decline with age. Carbon storage is maximized in old forests. Maintaining forests with multiple age classes of trees will provide a balance of large, older trees for storage and younger, faster-growing trees for sequestration. In addition, multi-aged forests increase a forest's resilience to natural disturbances.

Trees of different ages often vary in height, which increases the diversity of vertical structure within the forest. Forests with multiple layers will store more carbon. Implementing strategies that allow for the development of a multi-aged, vertically stratified forest will provide the opportunity to increase levels of "carbon packing."

Consider Species Composition

Achieve a vigorous forest by establishing and promoting native, locally adapted tree species that have no known forest-health issues and that are predicted to be competitive in future climate conditions—especially drought. Encouraging a diversity of species will also increase the forest's resilience to natural disturbances by ensuring that diseases or insects that kill one species will not kill an entire forest. Other factors to discuss with your forester might include the following:

- Promoting trees such as red oak and white and red pine, which have the capacity to become dominant and grow very large, can increase carbon storage.
- Tree species have different wood densities. Promoting tree species with high-density wood that can grow to be dominant trees can increase carbon storage in a forest. For example, hardwood trees are denser than softwood trees. There are even differences among hardwood species. Red oak and sugar maple are denser than red maple.
- Promoting shade-tolerant trees such as sugar maple, which can grow in the shade below the main canopy, can help increase the number of live trees growing in the forest, maximizing the opportunity for carbon packing by creating forests of multiple layers.

Deadwood Pool

As mentioned earlier, designating retention trees to leave standing in your live aboveground carbon pool will ensure a future source of deadwood, as the trees are left on-site until they die. Here are some additional strategies for maintaining your deadwood pool:

- Work with a forester to maximize the amount of slash left on-site after a timber harvest. Be sure these standards are included in your logger's contract.
- Felling or girdling poor-quality trees will add to the deadwood pool while also providing habitat benefits and freeing up space and resources to increase the growth rates of adjacent trees.

USING THE PASSIVE AND ACTIVE APPROACH TOGETHER

It doesn't have to be all or nothing! When considering the trade-offs of your decisions, it is important to realize that you do not need to treat your whole property in the same manner. You can decide to engage in active forest management on some parts of your forest and not on others. In fact, by leaving retention trees within a harvest, you are taking both an active and a passive approach within the same area. Similarly, those making decisions on public or private conservation land do not need to treat each of their individual properties the same way.

Your forest is part of a complex patchwork across the landscape. Each property has certain characteristics that make it unique, and each is surrounded by other unique properties. Therefore, each forest should be considered individually and within the context of its surrounding landscape with the help of a qualified forester. Some properties may be better suited to active forest management, while other land may be better suited to a passive approach. You do not have to feel the pressure of meeting all the demands of our region's forests. Ultimately, we need landscapes with both active and passive approaches for the many benefits forests provide, including carbon storage and sequestration.

FIELDWORK

What Is Your Forest's Role in Sequestering and Storing Carbon?

Refer to the illustration on page 141 to identify the carbon pools within your forest. What stage(s) of forest development is your forest in and how does that relate to its ability to sequester and store carbon?

How does promoting carbon sequestration and storage complement your landowner goals?

Does your forest have simple structure, or is it complex, with diverse ages and sizes of trees? What are the implications of this structure in relation to forest carbon?

Do you have a formal plan for the future of your land that will keep the forest as forest, maintaining its ability to sequester and store carbon?

Practicing Ecological Forestry to Mitigate Climate Change

CONTINUITY: The aboveground carbon pool is one of the largest pools in the forest. A disproportionate amount of aboveground carbon is stored in the largest-diameter trees. These trees take decades and even centuries to develop. Identify individual large-diameter trees and patches of large trees to retain during a harvest to reduce the loss of carbon from the forest. Of course, retaining large trees can represent the loss of potential income from their sale, a trade-off that you'll have to consider. The soil carbon pool is the other large carbon pool in the forest. Maintain soil integrity through the application of best management practices (BMPs), robust timber harvesting contracts, and harvest supervision.

COMPLEXITY AND DIVERSITY: Forests with high structural complexity have higher rates of carbon storage than those with simple structure. Encourage high structural complexity, including vertical diversity and high quantities of deadwood. Be sure that challenges such as invasive plants and excessive herbivory aren't simplifying forest structure and limiting tree regeneration.

TIMING: Extend the time between regeneration harvests to allow forests time to sequester and store additional carbon. For example, delay the harvest for 10 to 20 years or wait until the trees have added 2 to 3 inches of diameter.

CONTEXT: A diversity of forest types and age classes will help balance carbon sequestration and storage. This diversity will also create "resilient carbon" by ensuring that the loss of carbon from a large, single disturbance will be minimized by having a diversity of species and ages, and also allow for a faster recovery of carbon stocks post-disturbance.

CASE STUDY

Northam Forest Carbon, Jockey Hill Farm

Tim Stout

LOCATION: Shrewsbury, Vermont

ACRES: 173

A retired executive of a Boston-based utility company, Tim Stout spent much of his professional career focused on designing and implementing large-scale energy efficiency programs for residential, commercial, and industrial customers. It is not surprising, therefore, that when he heard a presentation on the important role forest carbon plays in mitigating climate change he saw how his professional interest in climate

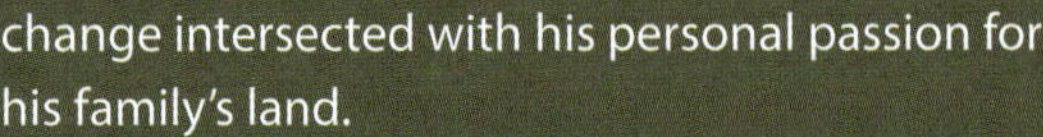

change intersected with his personal passion for his family's land.

Ever since, Tim has worked to steward his forest to increase its carbon benefit through climate-adaptive management strategies, such as planting climate-resilient tree species, eradicating invasive plants, and being an early adopter of the American Forest Foundation's Family Forest Carbon Program. In early 2024, Tim joined the board of the American Forest Foundation.

Tim's impact extends well beyond the boundaries of his valley. His passion for addressing climate change led him to start an organization called Northam Forest Carbon (northamforestcarbon.com). The mission of Northam Forest Carbon is to "educate landowners about the fundamentals of carbon sequestration and storage and encourage and facilitate landowners to adopt management practices that will optimize carbon storage and sequestration in order to reduce the carbon in the atmosphere." Tim achieves this mission through a variety of strategies, including hosting landowner tours on his property, walking the forests of other landowners, and developing resources, including a poster describing the ways family forest owners can help mitigate climate change.

To extend his reach and complement the work of other organizations, Tim works

In every decision, the next generation of landowners is always in mind.

with Vermont Coverts, Vermont Woodlands Association, and several other organizations. Tim also uses drone photography to help other landowners see their land in a new way, by providing aerial photographs of their land. The aerial photographs help landowners envision their land and its context, see the forest carbon on their property, and start conversations about ways to increase the carbon within their forest and engage in legacy planning to ensure the land will stay forested in the future.

Beyond Tim's family's land, there are several other privately owned forests in his valley. In addition, the Calvin Coolidge State Forest abuts this block of private landowners. In total, the private landowners and state forest comprise 2,300 acres within the valley. Tim has cultivated relationships with each of his neighbors, including longtime residents and new landowners. The collective relationship he has fostered with his neighbors is the foundation on which their landscape-scale forest management is built, as Tim provides information and organizational capacity to coordinate efforts, including joint timber harvests and a landscape-wide effort to remove invasive plants. Tim even shared his bird survey done by an ornithologist friend (conducted over a three-year period) with the management foresters at the local state forest to help them better understand current wildlife populations.

These collaborations give Tim hope that, if he were to pass away tomorrow, there are enough people in the family as well as outside the family who could carry on the legacy of forest management. "Clearly, more needs to be done. But if we get more and more of those people together, we can strive for longer-term goals," he says.

Forest legacy is of great importance to Tim. The land was originally purchased by his great-grandfather in 1942. Since that time, there have been six generations of his family stewarding the land. The stewardship responsibility now falls to Tim and his brother. They have put formal conservation-based estate plans into place to ensure the transition of the land to the next generation. These formal plans are also complemented by efforts to connect future generations to the land and their family's traditions. In every decision Tim makes, the next generation of owners is always in mind.

When asked about the advice he would pass on to other landowners, Tim points out that all he has achieved is a function of his network of professionals and friends. He has found that all the professionals with whom he has spoken have been generous with their time and expertise. "The level of collaboration among all these parties has been truly inspiring," says Tim. In addition, Tim believes that meeting other landowners is the most rewarding part of his work: "Hearing the rich stories of other landowners who are finding their passion for their forest is very rewarding. We always learn from one another."

"If you build it, they will come."
—*Field of Dreams*

CHAPTER 8

Diversifying Wildlife Habitat

WE OFTEN ARE ATTRACTED TO THE BIG AND BEAUTIFUL ELEMENTS of the forest, including wildlife. Seeing a deer run through the woods, bounding over a stone wall with its white tail flying, is majestic. Watching a bear lumber among the trees, searching with its nose for food, is humbling—and a little intimidating. Hearing an owl is chilling and magical. Seeing a woodcock doing its mating dance is mesmerizing. Witnessing a beaver working to create its habitat is impressive. These kinds of experiences encourage us to help sustain the many species with whom we share our forests. The more we interact with wildlife, the more connected we feel.

There is something special about seeing wildlife on your land and a sense of pride that comes from knowing that your forest is contributing to the survival of wild species. Supporting wildlife and all members of the forest community is part of a reciprocal relationship to the forest. Hosting wildlife is a way to give back.

A male ruffed grouse stands on a drumming log to claim his territory and attract a mate.

Black bear populations have increased as the amount of mature forest has increased.

WHAT IS WILDLIFE?

When using the term *wildlife*, it is common for people to refer primarily to vertebrate species, especially mammals, birds, and fish. Wildlife serves as an indicator that, despite a period of intensive land use, our forests and the associated species that rely on them are coming back. Any management strategy you implement, even a passive one, will benefit some species and discourage others. You may have particular species you are interested in, or you may want to support a broad diversity of species. Either way, ecological forestry can help create habitats and support the full range of biodiversity on your land.

Some landowners have very specific interests in certain kinds of wildlife. Those who hunt may want to help increase the population of game species such as turkey, deer, grouse, and bear. Landowners interested in birding may seek to provide habitat for particular bird species. Recent reports on the crash of bee populations and the dire ramifications of this decline have inspired some landowners to give more attention to the world of invertebrates, providing habitat and forage opportunities for pollinators to help reverse their decline.

On the other hand, you may have a natural and understandable reluctance to embrace particular species that are seen as pests. For landowners engaged in agriculture, wildlife can cause damage to crops, orchards, sap lines, and berry fields. Animals such as racoons, skunks, snakes, rodents, and insects are not likely to make the top 10 list of favorite landowner species. In fact, these species may actually strike fear into the hearts of some. In general, we prefer warm and fuzzy wildlife species that don't eat our food or live in our homes. This bias for certain species over others can influence our goals and forest management decisions.

FIELDWORK

What Are Your Wildlife Interests?

Do you have certain species that you are interested in supporting, or is your interest in wildlife more general? Are you more interested in the habitat provided by the preforest, young forest, mature forest, or old forest part of the successional clock, or some combination? Do you have any unique habitats on your land, such as vernal pools, wetlands, streams, underrepresented forest types, or rocky outcrops? Is your land representative of your larger landscape or is it different?

Many landowners don't have a particular species in mind at all when they set out to pursue wildlife goals; they just have an interest in providing habitat. If this describes you, then you can choose to contribute to wildlife in a broader way by developing strategies that make the most of the unique aspects of your land and surrounding properties. Regardless of your interests, there are a few basic things on which all wildlife depend—and which you can help provide.

WHAT IS BIODIVERSITY?

Over the last couple of decades, there has been increasing interest in expanding the focus of wildlife management to promote biodiversity, the diversity of all life. Rather than focusing only on vertebrate wildlife species, especially mammals and birds, biodiversity is meant to widen the focus to include the variety of all living organisms, from the smallest fungi in the soil to apex predators and everything in between.

When we broaden the focus from wildlife to biodiversity, we start viewing the forest for all of its parts, not just the biggest, most beautiful, or most useful. At its essence, ecological forestry is focused on restoring and maintaining a forest's ability to function fully. Properly functioning forests are able to regenerate, grow, develop, evolve, and recycle. Achieving this takes the work of millions of species, many of which we never see because their role in the forest ecosystem plays out underground or they are microscopic in size. Nonetheless, these are the ecosystem engines, the drivers of change and resilience. Though harder to measure and therefore harder to track through time, there are opportunities to steward a forest in a way that is not only beneficial to wildlife but also purposeful in its protection and enhancement of biodiversity.

One example of ecosystem engines that are essential to forests but not usually thought of as wildlife are fungi. Fungi are spore-producing organisms that feed on organic matter: deadwood, branches, leaves, needles. They break down that organic matter, returning the nutrients back to the soil to be recycled by current and future trees. Another critical contribution of fungi is in assisting plants and trees to access water and nutrients. Mycorrhizal fungi are found in the soil and have a symbiotic relationship with trees. A symbiotic relationship is one in which both individuals in the relationship benefit the other. In this case, the tree provides food in the form of sugars to the mycorrhizal fungi that grow among its roots. In return, the mycorrhizal fungi increase the reach of the roots, giving them more surface area with which to absorb water and nutrients that are used by the tree. This symbiotic relationship plays a very important role in tree survival, especially when

Ecological forestry seeks to preserve the full range of biodiversity in order to nurture all the parts and processes of the forest. Pictured here are tree lungwort and a bumblebee.

trees are seedlings and vulnerable to droughts.

Lichens are another important organism for forest biodiversity. Lichens are really composite organisms comprised of symbiotic relationships between fungi and algae. Lichens can grow on rocks and trees. Because lichens are slow-growing, the presence of certain lichens on trees is an indicator of an old tree. Tree size and tree age do not always correlate. While we can grow big trees relatively quickly through crown thinning, that wouldn't accelerate the development of lichens on the tree. While crown thinning can make a tree larger, faster, it can't make it older. It takes time for lichen and other species, such as understory plants, to develop. Sometimes good things just require time, another ecological forestry principle.

Insects also play an important role in biodiversity. Critical to agriculture, bees are most often thought of as an open habitat or meadow species. However, there are dozens of bee species in our region that are forest dependent. Deforestation of our landscape due to agricultural practices likely resulted in an undocumented loss of forest-dependent bee species. Despite this loss, a number of forest-dependent bee species remain in our landscape. Some forage on understory vegetation such as spring ephemerals and other plant communities often found in old forests. Creating patch reserves, as described in Chapter 6, helps provide opportunities for these types of plant communities to develop and serve as a seed source for repopulating forest areas across the landscape. Though harder to identify and quantify, some forest-dependent bee species use the flowers of trees.

Let's also not forget that disturbances are a natural part of forest development and that our region's disturbances tend to be frequent and low intensity, with occasional larger disturbances. These disturbances create patches of young forest that can include understory plants similar to those found in open habitats in which bees often forage. In a dynamic landscape, disturbances form a shifting mosaic of understory plants to support a diverse suite of pollinators, reinforcing the importance of multiple age classes within the forest and on the landscape.

WILDLIFE HABITAT BASICS

Habitat is a defined area within a landscape that includes all of the needs of a wildlife species. Your land may have many habitats beyond forests, including agricultural fields (crop, pasture, hay), orchards, water resources (wetlands, streams, ponds/lakes), and other unique areas such as power line rights-of-way. So, while forest habitat may logically be the first one you think of when talking with your forester, the reality is that you should consider your forest as part of a larger whole. Think about your land as an integrated series of different ecological parts.

Likewise, your land is part of a patchwork quilt of other lands, each with their unique qualities. This landscape perspective is important in all aspects of ecological forest management, but it may be most critical when considering wildlife. When seeking to find habitat to satisfy their needs, wildlife species don't start and stop at human property boundaries. Instead, they utilize all the ownerships within their range. Your land may provide some or all of the particular food, water, or shelter for a species, depending on the species and the size of their range. It is more likely that your land is providing part of the need.

When evaluating the landscape, you'll need to consider both the extent to which your land is similar to the surrounding properties and how your land is connected to similar habitats on other properties. Is your land providing something special to the landscape, such as forest of a different age or species composition? If so, how can you maintain that benefit? If not, can you change what your land is providing or add to a benefit that is lacking within the landscape? To evaluate this, it's important to have a clear understanding of species' needs.

Forest Structure as Habitat

Like all living things, wildlife needs a place to live that provides food, safety, protection from predators, and the opportunity to raise their young. These needs may be met through one type of habitat, but most often wildlife uses different types of habitats to meet these needs. Approximately 20 percent of vertebrates rely especially on one stage of forest development, the most common being preforest. The remaining 80 percent of vertebrates require a variety of stages within the spectrum of forest development to meet the various needs of their life cycles. This is because wildlife species have evolved over time to use different stages of forest development created by regular disturbances.

Early wildlife biologists adopted forestry terms to describe the various ages of forests. Since the origin of forestry in the US was primarily focused on wood products, forest size classes are described using forest products terms. Despite their history, these terms continue to be useful:

- **Seedling** forests are the result of a significant disturbance, either natural or human-induced, in which most of the forest was removed. As the name suggests, these forests are composed of small trees shorter than 4.5 feet tall (breast height).
- **Saplings** are the next size class and include trees that are taller than breast height and up to 4.9 inches in diameter at breast height (DBH).
- Once trees reach 5 inches DBH, they are no longer considered part of the regeneration layer and are referred to as pole-timber. Trees remain pole-timber until they grow to approximately 11 or 12 inches DBH.
- Trees larger than 11 to 12 inches DBH are considered sawtimber. Within the sawtimber class there are typically three sizes: small sawtimber (12 to 15.9 inches DBH), medium sawtimber (16 to 19.9 inches DBH), and large sawtimber (20 inches DBH and larger).

Though many of the forests in our region are even-aged, meaning they started after a large disturbance event (abandoned pastures reverting back to forest or an old clear-cut regrowing, for example), they are increasingly taking on a more complex structure the older they become and the more they are influenced by disturbances. Forests with three or more age classes within them are called uneven-aged forests, in recognition that the trees within the forest were initiated at different times.

With these size classifications in mind, it is possible to consider the size class(es) creating the structure of your forest. The land use history of the property is often a significant determinant of the structure of the forest. If the property supported agriculture and was abandoned 50 to 80 years ago, then it's likely that the property is in sawtimber. Not all parts of the forest have the same land use history, though. It would be common for areas of the farm farthest from the barn to be abandoned first and areas closest to the barn and homestead to be abandoned last. Of course, the structure of your forest is not only the result of human land use history but natural disturbances as well.

Another way to characterize forested habitats is by using the forest stages that we described in Chapter 1: preforest, young forest, mature forest, and old forest. Preforest, created after a significant disturbance, includes herbaceous plants and tree seedlings. As the forest grows into saplings and poles, the forest has entered into the

FIELDWORK

What Is Your Forest's Structure Like?

Perhaps the biggest driver of forest habitat is its structure. Walk through your forest and determine the stages of development. Is your forest all in the same stage of forest development or are there different stands in different stages?

Also think about the structure of other forests in your landscape. Start with your immediate abutters. Are their forests similar or different to your forest in structure? Move out beyond your direct abutters. How would you characterize the forest structure of the landscape? Are there any forests owned by the public or conservation organizations for which you have a sense of their possible future forest management based on their mission? Does your land serve as an important connection to other forested areas in the landscape?

Remember, many of our forests are similar in age, therefore diversifying habitats is often about either moving the hands of the successional clock back to create preforest or moving them forward (passively or actively) to create old forest. Of the forested area in the landscape, roughly estimate the proportions of preforest and old forest stages. What role is your land playing in the landscape? What role could it play?

young forest stage. Forests in the mature forest stage include trees in the sawtimber size classes. Older and larger trees can be found in the old forest stage, which also includes a diversity of tree sizes and ages.

Regardless of the way in which the structure of your forest is characterized, your goal is to gain a sense of the type of habitat your forest has so that you can understand the types of species likely using your land. Then you can make a decision as to whether you would like to maintain the habitat or diversify it through the creation of different forest structure. However, your land is only one part of most species' habitat, so to see the whole picture, it's important to consider how the habitat on your land relates to the habitats in the surrounding landscape.

LANDSCAPE CONTEXT FOR WILDLIFE

The parcel sizes of family-owned forest properties are most often in the range of dozens of acres to a few hundred acres. As such, no one property serves as the habitat for a species' entire life cycle. Consider how much of the land in your landscape is forested. Also evaluate the characteristics of your forest relative to the other forests in the landscape. Is your forest unique in its structure or species composition? If it is unique, then maintaining it may help provide diversity within the landscape. If your forest is similar to other forested portions of the landscape, there may be opportunities to diversify the structure within the landscape.

Does your property connect to the other forests, perhaps through contiguous forest cover or a stream? If so, are there plentiful connections between other forests or is yours one of the only ones? If your forest is one of the few ways in which an animal can move through the landscape, there is good reason to maintain the connection with continuous cover. If there are other ways for species to move through the forest, then there are more opportunities to diversify the habitat within the landscape. What is the habitat on the abutting properties? Though we want diverse habitats in the landscape, the habitats should be a minimum size. Generally speaking, the bigger and more unfragmented the habitat, the better.

Setting Achievable Goals

An important goal relative to wildlife habitat is articulating what type of habitat you want to create or maintain and in what amount. For many conservation organizations and public wildlife agencies, their broad goal is to maintain the full range of native species within the region. This is good and admirable, but sometimes the devil is in the details.

As we have discussed, our forests have gone through radical change in the last couple hundred years. In some places, our landscapes have gone from mostly forested to mostly pasture and back to mostly forested, whereas in other places, forests were never cleared but instead highly simplified by intensive harvesting. The end result in both cases are forests that are far more ecologically young than the original forests. So, when the goal is to conserve all native species, which era of forest history are we using as the benchmark, both to identify the species to conserve and to determine the target amount of habitat necessary to sustain the population at its historic level?

Rather than wade into the fray of this sometimes controversial question, it's more productive to consider the conservation steps proportionally appropriate for your forest. The reality is that most landowners don't own enough land to influence an entire population of wildlife species and achieve broader biodiversity goals. Instead, consider your personal goals and the way in which your land can be stewarded to maximize the overall impact at a landscape scale.

To support native biodiversity, there needs to be a diversity of habitats across the landscape—some preforest, some young forest, some mature forest, and some old forest. The fact that most of our forests are 60 to 100 years old is very good news for those species that depend on mature forest habitats, and, indeed, we have seen the return of many mature forest species, including black bear, moose, and fishers. However, since our forest ages are so uniform, our landscapes are often sparse in preforest and old forest stages, the latter having once been the predominant habitat across our region. This means that diversifying wildlife habitats often involves focusing on either moving the successional clock back to a preforest condition

FOREST MANAGEMENT PLANS

Assessing your forest's structure and species composition is an excellent way for you to engage with your forest and learn more about it. You can also contact a forester to walk your land and give you a general sense of your forest and its landscape context. For a formal and very detailed description of your forest and its resources, a consulting forester can be hired to develop a forest management plan based on field inventory and mapping. In general, a forest management plan describes your goals and your forest, making recommendations on strategies to achieve your goals given your current forest conditions. To be clear, you don't need a forest management plan to engage with your forest, but it can be a good starting point to help you understand your forest and forest management strategies to reach your goals.

A forest management plan typically is written for a 10-year window and serves as a road map to your forest. Having a forest management plan for your forest can be a requirement for qualifying for state and federal cost-share programs and state current use property tax programs. More information on current use programs can be found in Chapters 10 and 12.

or moving the clock forward to reach old forest conditions.

For those landowners with an interest in maintaining or creating habitat for a particular species or set of species, the decisions become relatively straightforward; it's a matter of assessing your land's ability to provide for the needs of that species.

FOOD SOURCES

Like all living things, wildlife must have food to sustain itself. Of course, the specific type of food is species dependent. The foundation of the food web is made up of green plants that produce their own food through photosynthesis, and these plants make up the diets of the ecosystem's herbivores. Carnivores eat the herbivores and other, smaller carnivores. Together these relationships among plants and wildlife make up the food web, with each relationship connecting species across the habitat.

Deer and moose are excellent examples of herbivores, as they depend completely on plants to meet their nutritional needs. Buds and the new, tender growth of seedlings and young trees provide forage for them. A deer must eat around 7 percent of its body weight in vegetation per day to remain healthy. For a 150-pound deer, that's equivalent to about 10 pounds. A moose must eat more like 50 pounds per day. To help meet their daily requirement in winter when other sources of browse are scarce, moose will also scrape the bark off of striped maple, also known as moose maple. The amount of forage necessary to sustain deer and moose can be absorbed by a forest when populations of the animals are balanced. However, when these populations increase, the impact on forest vegetation, particularly on tender regeneration and new growth, can be profound. More about the impacts of excessive herbivory can be found in Chapter 9.

Fruit trees, including remnant apple trees from past use, and fruit-bearing shrubs like blueberry also serve as attractive food sources for many species. Identifying these food resources on your land and liberating them from competing vegetation can improve their ability to produce.

Another important plant-based food source, particularly in the fall and winter, is mast. Mast is the fruit of trees and shrubs and can be categorized as either hard or soft. Acorns, hickory nuts, and beechnuts, with their tough exteriors, are called hard mast. Deer prefer white oak in the fall when they are gaining weight for winter, since it is higher in fat and lower in bitter-tasting tannins than red oak. Despite the bitter taste, red oak acorns with their high tannin levels are rot-resistant, making them a valuable food source into the spring when foraging sources are limited. Examples of soft mast include black cherries and white pine and hemlock seeds from cones. The fruits of black cherry and pin cherry are among the most valuable soft masts, as numerous species of wildlife feed on them.

Up the Food Chain

Species that feed on plants serve as a food source for a whole suite of wildlife species. Plant-eating insects, for example, are eaten by many species. Many birds are insectivores, sustaining themselves on the various protein-rich stages of an insect's life cycle from larva to adult. Some birds, such as swifts, swallows, and flycatchers, are aerial insectivores, catching insects midair in fields and open areas. Other birds forage insects off of various parts of the tree, including the stem of the tree and its branches. Large, old trees with complex bark and complex crowns, like those retained as biological legacies, can harbor a wealth of insects and support a diversity of birds.

It's not just live trees that host insects. Cavity trees—living and dead trees with areas of rot that form cavities—are home to a rich source of boring insects that feed on cellulose and the living cambium layer of bark. These insects are the target of woodpeckers. The woodpeckers' excavations, in turn, create holes in the trees that are then used by birds and small mammals as shelter. Beyond birds, species of mole salamander also depend on insects for at least part of their nutritional needs. Jefferson, marbled, and spotted salamanders spend most of their lives in burrows in upland forest and are voracious predators of insects. In wet areas, aquatic insects are also a food source for fish (such as brook trout), frogs, and salamanders.

At the top of the food chain are predators. Apex predators are charismatic species such as bears, mountain lions, wolves, and hawks. Predators aren't necessarily only carnivores. Many are omnivores, eating both vegetation and meat. A quick look at fox scat, for example, will often show the remnants of plant-based food (such as black cherry pits) as well as rodent hair.

The interconnection of plants and wildlife with other wildlife cannot be overemphasized. A common example is the boom-and-bust cycle of mast-producing

Acorns and other mast are important food sources for a variety of wildlife species.

Blue-spotted salamanders use wetlands and moist deciduous and mixed forests for habitat, breeding in vernal pools.

trees, such as oaks. An oak will not produce acorns every year, but rather every two to five years. In a mast year, many oaks in a landscape produce mast at the same time. The precise mechanisms by which trees are able to synchronize mast production—and why they do so—are not fully understood. Producing a lot at once improves the likelihood that the acorns will overwhelm the populations of those wildlife species that feed on them so that some acorns are left to germinate.

However, the implications go well beyond new oak trees. Increases in mast mean food for species that feed on acorns, particularly rodents (mice, squirrels, chipmunks). Their populations increase in response to the availability of more nutrition, and they in turn become food for predators, increasing their populations in the following year or two. Eventually, without acorns the rodent population is reduced, followed by the predator population. The cycle starts over the next mast year.

WATER SOURCES

Wildlife species also need to have access to water. We are very fortunate in our region to generally have ample water in a variety of forms. Perennial streams (those that run year-round), intermittent streams (those that run for only part of the year), wetlands of various types, vernal pools, ponds, and lakes can all be found on and around the forests of our region. Water features, particularly stream systems, are often surrounded by forest, connecting parts of the landscape and making them prime locations for wildlife species to find food. Amphibians breed in water. Birds and other species bathe and preen. Even seeps—areas on a hillside where water hits an impervious surface, like bedrock, and comes out of the ground—can have very important wildlife value as they are often the first place where snow melts and plants begin to grow in spring, providing food after a long winter.

Ecological forestry strategies for water resources focus more on protection than on active management to enhance them. However, restoration of in-stream deadwood has become an important strategy in some regions to support fish populations and reduce flooding impacts.

When making decisions about any type of active forest management, be sure to pay close attention to protecting water resources. It's likely that your state has a series of voluntary or mandatory actions for protecting water resources. Examples include placing riparian or buffer strips along water resources. At the same time, if you're interested in restoration opportunities tied to increasing the complexity of aquatic habitats, namely streams, connect with a forester to learn more about how to best do this on your land.

FIELDWORK

What Food and Water Sources Are on Your Land?

Walk your land and identify current food sources. Do you have tree species that produce hard mast, such as acorns or other nuts? Do you have species that produce soft mast, such as black cherry, pin cherry, pine, or hemlock?

Do you have large, old trees that may host a diversity of birds and insects? How about cavity trees and downed deadwood?

What water features or other unique habitat features are on your land?

MANAGING FOR DIVERSITY

Active forest management can help establish and enhance food resources for wildlife. Encouraging a diversity of food requires attention to diversity of both tree species and age.

Create different light levels. Forest management can be used as a tool to increase the proportion of trees with high wildlife food value. Black cherry is a shade-intolerant species, meaning that it is most competitive in sun. Oaks are typically a mid-tolerant species, meaning they are most competitive in partial sun. Regenerating these species requires creating the appropriate light levels to favor them. Encouraging a variety of species out on the landscape requires creating different light levels.

Encourage mast production. It takes energy for trees to produce mast. Thinning around the crowns of individual trees producing desirable mast will allow them to increase their crown size and mast production.

Maintain unique species. When thinking about your forest's ability to host a diversity of wildlife species, consider the tree species growing in your forest and the extent to which they are represented in the surrounding landscape. In general, maintain or establish species not well represented in the landscape.

Create diverse age classes. A diversity of tree ages will provide a range of browse options, potential to host insect populations that serve as important food sources, and different types of hiding cover and foraging substrates. Patches of young trees will provide browse, insects, populations of rodents like mice and rabbits, and dense hiding cover. Conversely, patches of old forest will provide old trees with complex bark and crowns and abundant standing and downed wood. The plants and insects that use these various habitats will support many other species. Consider the extent to which each age class in your forest is represented in the surrounding landscape. In general, maintaining or establishing age classes not well represented in the landscape will help sustain a greater diversity of species.

Identify unique habitats and resources. These areas are often associated with water resources, such as vernal pools, wetlands, and streams. Other areas of high value include rock outcrops, patches of old forest,

FIELDWORK

What Is the Species Composition of Your Forest?

Looking at species composition can involve a very general or a detailed assessment. At the most general level, is your forest comprised mostly of deciduous trees (those losing their leaves in the winter) or conifers (those maintaining their needles throughout the winter)?

If you're interested in greater detail, consider the forest type(s) in your forest. Is it northern hardwood (beech-birch-maple), spruce-fir, oak-pine, oak-hickory, aspen-birch, or something else? Now consider the species composition in the landscape. Does your forest offer a diversity of species? Is the species composition of your forest unique or is it similar to that of surrounding forests?

areas of preforest or grasslands, and unusual understory plant communities. In general, unique environments often support unique species that have evolved adaptations to be competitive in these environments, so also consider areas that are very high in the landscape and very low as well as areas that are very dry and very wet. To the extent possible, identify these same areas within the landscape. Some may be easier to identify than others.

Consider other forms of biodiversity. Fungi play an essential role within forests by breaking down woody organic material and recycling the nutrients, adding to the soil's fertility and carbon content. Different types of fungi prefer different species of wood. Different fungi also prefer wood in different states of decay. Different species of trees in various stages of decay will support a broad diversity of fungi within the forest, helping to ensure the cycles of the forest are fully functioning. Making sure your forest has appropriate levels of deadwood in various states of decay helps achieve forest continuity and complexity, leading to increased levels of biodiversity.

Dead trees of diverse species in various states of decay support a wide range of fungi and other organisms.

CASE STUDY

Building a Future for Both People and Wildlife

Alan Finifrock

LOCATION: Nemadji and Barnum, Minnesota

ACREAGE: 160 acres and 120 acres

Alan and his wife, Sharon, own two properties in Central Minnesota, southwest of Duluth. The original family homestead in Nemadji is enrolled in the Sustainable Forest Incentive Act, Minnesota's current program that provides property tax relief in exchange for implementing sustainable forest management practices. The second property, in Barnum, with frontage on Bear Lake, has a conservation easement on the land held by the Minnesota Land Trust.

Alan credits his parents, who labored on the farm he now stewards, for instilling within him a land ethic. Aldo Leopold was also a significant influence. Although Alan manages the land as a business, he describes his relationship to the land as a spiritual one: "I believe I am a custodian of the land and that the land really belongs to God. The land is a gift to me, I'll say, from God, but through my parents." Alan has worked to share the gift of his land with his family and many others to sustain the land's gifts.

As a former teacher, Alan has sown the seeds of a land ethic among many young people. He has also planted about 30,000 trees, most of them by hand. In the mid-1970s, Alan began annual plantings of red pine and white spruce seedlings. Over the years, he has recruited about 25 young people from his family and the community to help with the plantings. Several of those kids have returned to the property to see the results of their efforts, and Alan has pictures of them standing next to the trees they planted. Both trees and kids have grown and matured, but they remain connected. Planting is most certainly an act of faith in the future—faith in both forests and people.

Alan has placed a significant focus on supporting wildlife on his land. With the help of many professional foresters and biologists, he has developed stewardship activities that recognize the importance of structural complexity within the forest and the importance of age class diversity within the forested landscape. Over the years, he has implemented several timber harvests to regenerate aspen for grouse habitat, built a wildlife pond, implemented prescribed burns, and managed the forest to support golden-winged warblers, a species designated as near threatened. Alan works with the American Bird Conservancy to monitor the presence of neotropical migrants.

Planting trees is an act of faith in the future—faith in both forests and people.

Although the golden-winged warbler is closely associated with young forest and shrub habitat for nesting, they need a diversity of forest structure to complete their life cycle and raise their young. On Alan's land, timber harvesting was used to create the conditions necessary for ground nesting, while also leaving small patches to preserve a mid-level canopy for feeding and tall trees within the stand from which the birds can sing and protect their territory. The very next year after implementing these practices, there were golden-winged warblers using the habitat. These management strategies demonstrate the importance of the ecological principles of continuity by retaining trees and patches, complexity by encouraging diverse forest structure, and context by considering the land's role within the landscape.

Of course, forest continuity also has a social context. Alan and his wife have been very intentional about connecting the next generation to their forest. For example, his family invented a game called "Bumper Ball," a form of baseball played in a dense Norway pine plantation. The batter would hit a tennis ball, which would bounce off trees, pinball-like, while the batter ran the bases. Catching a fly ball made you the batter. Children, nieces, nephews, and foster kids all played Bumper Ball. Even decades later, Alan says they "still roll in laughter over the fun we had playing Bumper Ball."

Alan reminds us of the importance of not only our relationship with the forest but also our relationship with those around us. His advice: "Think long-term and involve family. It will be both fun and fruitful."

GETTING STARTED

Examining your forest's combination of structure, species composition, and unique habitats, as well as its landscape context, you will gain an excellent understanding of the habitats you are hosting within your land and those in the landscape and, in turn, the wildlife using these habitats. Then you can make decisions about how to support wildlife and biodiversity.

If the mix of forest stages, species composition, and landscape characteristics is supporting the type and amount of wildlife to meet your goals, there may be little need for intervention. However, if conditions are less favorable or if you would like to know more about your options, you can start by considering the amount of area in preforest and the amount in old forest. Consider opportunities to contribute to either of these stages of forest development. Pay particular attention to your forest's landscape position. If your land abuts preforest, your efforts to add to the habitat will increase its wildlife value. Likewise, if your land is embedded within a landscape of mature forest, including landowners likely to take the passive forest management approach, you could consider focusing on old forest development. Of course, if the proportion of preforest and old forest is already ideal, then diversifying the habitats in the landscape may provide the greatest wildlife and biodiversity impact.

Working with a forester is an excellent way to understand the characteristics of your land, the landscape, and your wildlife and biodiversity forest management opportunities. In fact, a wide range of programs specifically exist for landowners to work with foresters to enhance the quality of wildlife habitat on their land. There are many opportunities to build it and sustain it, if you have the interest.

Practicing Ecological Forestry to Diversify Wildlife Habitat and Promote Biodiversity

CONTINUITY: Some important wildlife characteristics can take decades and even centuries to develop, including large-diameter trees, snags, and large downed logs. Locate trees with important wildlife characteristics and retain them during harvests. Identify and protect unique wildlife habitats that may have a disproportionate amount of biodiversity, including water resources, rock outcrops, and well-developed understory plant communities. Retain trees producing important hard and soft mast. Leave all deadwood on site during a harvest.

COMPLEXITY AND DIVERSITY: Structurally complex forests are "niche-rich," supporting a wide diversity of habitats—understory, mid-story, overstory, deadwood. Facilitate the development of these layers. Deadwood, both standing snags and downed dead logs, support many wildlife species. Work to achieve appropriate levels of deadwood, and plan for periodic inputs to sustain those levels. Thin around individuals providing important mast crops to increase their production.

TIMING: Some wildlife and biodiversity characteristics can only develop through time. Examples include decay in living trees that may develop into cavities, the growth of lichens and fungi, and the development of complex bark. Designating legacy trees and patch reserves with the intention of never harvesting them will give them the time to develop important wildlife characteristics. Extend the time between harvests.

CONTEXT: Wildlife must move around the landscape to meet its needs. Diverse forest stages and well-connected habitats will support a diversity of wildlife by encouraging that movement.

> "It is not the strongest of the species that survive, nor the most intelligent, but the one most responsive to change."
>
> —Charles Darwin

CHAPTER 9

Building Forest Resilience

OUR FORESTS ARE INHERENTLY RESILIENT. As outlined in Chapters 2 and 3, they have endured extensive clearing, harvesting, disease, and other threats, and yet still we continue to enjoy the many benefits they provide. Today our forests face an uncertain future of rising challenges, such as forest conversion; invasive plants, insects, and pathogens; heavy deer browse; and climate change. These challenges are happening at a pace and scale never before encountered in the evolution of our forests.

The truth is, we don't know exactly how all of this will be sorted out, but we can use ecological forestry to prepare for this novel future by restoring full function back to our forests to maximize their ability to grow, change, and adapt and be as resilient as possible.

Our colonial land use history has resulted in homogenous landscapes comprised of forests with a similar age and species composition. Landscapes comprised of forests of diverse species and ages are more resilient to disturbances.

WHY FOCUS ON RESILIENCE?

Change in forests is natural and healthy, yet it is likely that we are at a time when the number of stressors facing our forests is greater than ever before. It's important to draw a distinction between the role natural disturbances play in forests and the impact of the challenges our forests now face. Natural disturbances foster complex forest structure by creating both more deadwood and gaps in the canopy that initiate forest regeneration. Natural disturbances are the driver of forest succession and development, and our forests have evolved strategies in response to them. While the current challenges facing our forests do produce some outcomes similar to those created by natural disturbances, such as gaps in the canopy and deadwood inputs, these challenges have some important differences. For one, the pace at which new challenges (such as invasive insects and plants) are moving, is likely outpacing the rate of past challenges, testing the forests' inherent ability to change and adapt.

The new challenges also operate at a different scale. Climate change is creating more frequent and more intense weather events. Invasive insects such as the emerald ash borer have led to tremendous mortality among ash trees across a broad region in a relatively short time. Finally, rather than disturbances leading to greater structural complexity, some of these new challenges result in forest simplification. Excessive deer herbivory can greatly reduce and even eliminate regeneration. Likewise, invasive plants can restrict regeneration and dominate forest understories, leading to structurally simple forests that are unable to replace the overstory trees when they die. This retrogressive succession reduces structural complexity and species diversity, leaving forests vulnerable to the challenges they face and with reduced capacity to store and sequester carbon.

Forests will continue to change in response to their environment, as they always have. Presumably, in the face of these changes, our forests will achieve equilibrium within a new normal. However, the length of time this will take is unknown. Trees are long-lived, and change is likely to take centuries, not decades. Meanwhile, challenges continue to arise. The process of evolution and adaptation is never-ending.

As we've talked about, different disturbances have greater impact on different parts of the forest. Therefore, building forest resilience focuses on encouraging complexity and diversity within the forests to help them adapt and be prepared for changes to come, despite uncertainty. By doing so, we are not putting all of our eggs in one basket. Instead, we are spreading out the risk so that when a disturbance comes through the landscape, the impact will be moderated. Like all the benefits we have discussed, increasing resilience by encouraging diversity and complexity must be done at both the property and landscape scales.

Working to increase your forest's resilience is an act of reciprocity in that it gives your forest the ability to fully express itself. Efforts to address landscape vulnerabilities can also lead to opportunities to connect with neighbors for greater impact. Restoring missing parts to the forest and preserving at-risk species also increases the resilience of Indigenous peoples and their cultural lifeways that depend on a full suite of forest species. Forest resilience exemplifies the connection between people and forests. To sustain one we must sustain both. Sustaining both requires a cultural shift in our relationship with forests.

Forest resilience: The capacity of a forest to respond to a disturbance by resisting damage or stress and recovering quickly. This can be measured in purely ecological terms (for example, do plants grow back quickly in disturbed areas?) or in the context of the values and benefits tied to forests (for example, do plants supporting the wildlife species we care about or a particular cultural lifeway grow back quickly?). When actively engaging with your forest to increase its resilience, you will work with its ecology, but your goals and actions will be generally guided by the values and benefits you hope to sustain into the future.

Forest stressors: Pressures on a forest that can reduce its resilience and impair its ability to function properly (such as drought, wildfire, or invasive insects).

Forest vulnerability: The extent to which a forest is susceptible to forest stressors.

FOREST RESILIENCE GRADIENT

Forest resilience can be thought of on a gradient. On one end of the gradient are those forests that have high resilience and low vulnerability. On the other end are those with low resilience and high vulnerability. Each property falls somewhere on the gradient based on its particular characteristics and those of its landscape.

High Resilience LOW VULNERABILITY	Low Resilience HIGH VULNERABILITY

FOREST RESILIENCE AT THE PROPERTY LEVEL

Although your forest faces an increasing number of stressors, the good news is that it may already have some resilient characteristics among the main ones outlined here. Understanding the characteristics of your land that give it resilience will help you better understand the ways in which you can sustain or even increase its resilience.

Increasing a forest's species and structural complexity will increase its resilience to forest disturbances.

Formal Plans for the Future of the Property

Most of the land in our region is owned by individuals and families. The average age of family forest owners is over sixty. This means that the coming years will mark the largest intergenerational transfer of land and assets our country has ever seen. The decisions family forest owners make (or don't make) about who will own their land when they are gone and how it will be used (such as converting it to another land use, parceling it into smaller properties, or protecting it from development) are likely the biggest and most permanent drivers of forest change we face.

The ecological forestry principle of continuity includes both ecological and social continuity. A resilient forest is one that will continue to be forest into the future. To ensure a critical level of forest cover, it is imperative that family forest owners make formal plans for the future of their land. Chapter 12 provides more information about strategies to develop a formal plan for the future ownership and use of your land.

Minimal Forest Stress

Each landscape has a unique combination of stressors and levels of exposure to them. Minimizing the presence of invasive species, lowering deer populations, ensuring that soil is abundant in organic matter and not compacted or eroding, and ensuring that water resources have forested buffers all minimize the number of stressors a forest faces and increase its resilience.

High Forest Complexity

The complexity of a forest is generally based on the following characteristics: tree species diversity, climate and disturbance suitability of tree species, tree size and age, tree arrangement, and the amount of deadwood present. Promoting characteristics that give forests complex structure helps them withstand and recover from stressors. A description of each of these characteristics follows.

Diversity of Tree Species

A forest is made up of different tree and plant species, which influences its ability to cope with stress and change. The two primary determinants of the trees and plants in your forest today are soil conditions (water and nutrients) and the length of time from the last major disturbance. Forests primarily made up of species better adapted to cool climate conditions or serving as host species for an invasive organism may be at greater risk. In contrast, forests that have a greater diversity of species are more resilient to many types of stressors simply because not all species are susceptible to the same challenges. Taking actions to increase diversity or encourage species that are likely to do better in the future can increase the resilience of your forest.

Ample Tree Regeneration of Future-Adapted Species

In addition to supporting a diversity of species, it is very important to consider maintaining and promoting species that are predicted to be well adapted—that is, more competitive—given expected future conditions. Thus, it is critical to consider how well a species is likely to do in a particular place, both now and in the future. Soil conditions (water and nutrients) play an important role in determining not only the tree species that can survive on a site but also those species that can thrive. Some trees compete very well on wet, nutrient-rich sites, while others compete best on dry, nutrient-poor sites. Some trees have a narrow range of soil conditions in which they will compete, and others will compete in a wide range of conditions.

Changes in temperature, evaporation, and precipitation are predicted to change soil conditions in our region. By varying the timing of timber harvests and creating different sizes of openings, trees that are better suited to the likely future climate and forest health issues can be regenerated, which will improve the likelihood of a vigorous future forest.

Planting and sowing tree seed is a common practice in parts of our region; however, the main goals have historically been to regenerate commercially valuable species, like red pine or black walnut. In the context of building resilience, planting and sowing seed takes on a different focus. Depending on your goals and resources, there may be opportunities to use enrichment planting and seeding of native tree species to promote a forest with a species composition well suited to expected future conditions.

PREDICTED CHANGE IN SUITABLE HABITAT

The following tables provide tree species and predictions of how competitive they will be in the future under low and high greenhouse gas emission scenarios. The values following each species name indicate whether species-suitable habitats will increase (+), decrease (–), or stay the same (•) under projected climate change.

New England / New York includes Maine, New Hampshire, Vermont, Massachusetts, Connecticut, Rhode Island, and northern New York.

Mid-Atlantic includes southern New York, Pennsylvania, New Jersey, Maryland, Delaware, and Virginia.

Central Appalachians includes Ohio, West Virginia, and eastern Kentucky.

Central includes western Kentucky, Indiana, Illinois, and Missouri.

Upper Lake States include Minnesota, Michigan, and Wisconsin.

Based on Climate Change Tree Atlas, https://www.fs.usda.gov/nrs/atlas/tree

NEW ENGLAND/NEW YORK

SPECIES	LOW	HIGH
Balsam fir	–	–
Black spruce	–	–
Northern white cedar	–	–
Yellow birch	–	–
Balsam poplar	–	–
Tamarack	–	•
Red maple	•	•
Sugar maple	•	•
Eastern white pine	•	•
Paper birch	•	–
White spruce	•	•
Scarlet oak	•	+
Red pine	•	•
Pitch pine	•	•
Bitternut hickory	•	•
Northern red oak	+	+
Quaking aspen	+	+
Black cherry	+	+
Black birch	+	+
Black oak	+	+
Bigtooth aspen	+	+
White oak	+	+
American basswood	+	+
Pignut hickory	+	+
Shagbark hickory	+	+
Chestnut oak	+	+
THREATENED BY CURRENT FOREST HEALTH ISSUES (DO NOT TARGET)		
Black ash	+	+
Green ash	•	•
White ash	+	+
Eastern hemlock	•	•
American beech	•	•
American elm	•	+

MID-ATLANTIC

SPECIES	LOW	HIGH
Red maple	–	–
Eastern white pine	–	–
Quaking aspen	–	–
Yellow birch	–	–
Pitch pine	–	–
Red pine	–	–
Paper birch	–	–
Atlantic white cedar	–	–
Northern white cedar	–	–
Red spruce	–	–
Black spruce	–	–
Balsam fir	–	–
Sugar maple	+	–
Black cherry	+	–
Northern red oak	•	•
Loblolly pine	•	+
Sweet gum	•	+
Bitternut hickory	•	+
Virginia pine	•	•
Swamp white oak	•	•
Silver maple	•	•
Chestnut oak	+	+
White oak	+	+
Yellow poplar	+	+
Black oak	+	+
Black locust	+	+
Blackgum	+	+
Scarlet oak	+	+
American basswood	+	+
Black walnut	+	+
Pignut hickory	+	+
Shagbark hickory	+	+
THREATENED BY CURRENT FOREST HEALTH ISSUES (DO NOT TARGET)		
White ash	•	•
Green ash	•	•
Black ash	–	–
Eastern hemlock	–	–
American beech	•	•
American elm	+	+

CENTRAL APPALACHIANS

SPECIES	LOW	HIGH
Red maple	-	-
Black cherry	-	-
Black birch	-	-
American basswood	-	-
Bigtooth aspen	-	-
Yellow buckeye	-	-
Yellow birch	-	-
Red spruce	-	-
Quaking aspen	-	-
Swamp white oak	-	-
River birch	-	-
Red pine	-	-
Sugar maple	•	-
Yellow poplar	•	-
Chestnut oak	•	•
Black locust	•	•
Shagbark hickory	•	•
Eastern white pine	•	•
Pin oak	•	•
Pitch pine	•	•
White oak	+	+
Northern red oak	+	+
Black oak	+	+
Pignut hickory	+	+
Scarlet oak	+	•
Blackgum	+	+
Virginia pine	+	+
Mockernut hickory	+	+
Black walnut	+	+
Sycamore	+	+
Bitternut hickory	+	+
Loblolly pine	+	+
THREATENED BY CURRENT FOREST HEALTH ISSUES (DO NOT TARGET)		
White ash	•	•
Green ash	+	+
Black ash	-	-
Eastern hemlock	•	•
American beech	•	-
American elm	•	+

CENTRAL

SPECIES	LOW	HIGH
Yellow poplar	-	-
Shagbark hickory	-	-
Scarlet oak	-	-
Pignut hickory	-	-
Virginia pine	-	-
Eastern cottonwood	-	-
Eastern white pine	-	-
Paw paw	-	-
Ohio buckeye	-	-
River birch	-	-
American basswood	-	-
Bur oak	-	-
White oak	•	-
Black cherry	•	•
Northern red oak	•	•
Black walnut	•	•
Sugar maple	•	•
Silver maple	•	•
Chinkapin oak	•	•
Post oak	+	+
Shortleaf pine	+	+
Black hickory	+	+
Mockernut hickory	+	+
Sweet gum	+	+
Sycamore	+	+
Blackjack oak	+	+
Black locust	+	+
Hackberry	+	+
Blackgum	+	+
Bitternut hickory	+	+
Honey locust	+	+
Common persimmon	+	+
THREATENED BY CURRENT FOREST HEALTH ISSUES (DO NOT TARGET)		
White ash	+	+
Green ash	+	+
American beech	•	-
American elm	+	+

UPPER LAKE STATES

SPECIES	LOW	HIGH
Quaking aspen	-	-
Black spruce	-	-
Balsam fir	-	-
Sugar maple	-	-
Jack pine	-	-
White spruce	-	-
Balsam poplar	-	-
Red maple	•	+
Tamarack	•	•
Silver maple	•	+
Northern red oak	+	+
Bur oak	+	+
Eastern white pine	+	+
American basswood	+	+
Black cherry	+	+
White oak	+	+
Bigtooth aspen	+	+
Northern pin oak	+	+
Black oak	+	+
Bitternut hickory	+	+
Eastern red cedar	+	+
Shagbark hickory	+	+
Pin oak	+	+
Hackberry	+	+
Black locust	+	+
Black walnut	+	+
THREATENED BY CURRENT FOREST HEALTH ISSUES (DO NOT TARGET)		
White ash	+	+
Green ash	+	+
Black ash	•	•
Eastern hemlock	•	-
American beech	+	+
American elm	+	+

Vigorous Trees of Various Sizes and Ages

Some stressors have a disproportionate influence on trees of a particular size and age. For example, windstorms have a bigger influence on large trees with dominant crowns. Certain insects, such as white pine weevil and spruce budworm, are more lethal to trees of a particular size. Forests with a diversity of tree sizes and ages are more resilient to a given stressor, and there is a lower probability that any given stressor will affect a large proportion of the stand.

Trees in overcrowded forests compete with one another for limited sunlight, water, and nutrients. This severe competition can increase the stress of a tree and reduce its vigor, increasing its vulnerability to stressors. Reducing the number of trees in a forest can free up limited space and resources, increasing the overall vigor of the forest. In addition, in fire-adapted ecosystems, like pitch and jack pine barrens and red pine woodlands, reducing the number of trees can also restore key habitats and lower the risk of catastrophic wildfires in these forests.

Variety of Tree Arrangements

Complex forests also have variety in how trees are arranged spatially. This includes open areas associated with recent canopy gaps created by disturbance or stewardship actions, denser areas with multiple canopy layers, and other spots where there may be widely spaced older trees. A resilient forest has a range of these spatial conditions to create unique microsites (for example, gaps for regeneration, deeply shaded areas for habitat, and widely spaced trees that can withstand drought) that offer multiple ways to recover from and resist stress.

Appropriate Amount of Deadwood

One component that is scarce or lacking in many of our forests is deadwood—dead standing trees (snags) and large logs on the ground. Though this deadwood can make the forest look messy, it is a critical part of a healthy forest. Snags provide habitat for insects, birds, and small mammals. When snags or live trees fall to the ground, the logs provide habitat for another suite of species, including insects and amphibians. They also serve as "nurse logs" for certain tree species such as yellow birch, which regenerate well on these moist, organic substrates. Finally, logs can store carbon for decades.

Having an appropriate amount of snags and logs in the forest helps maintain these critical ecosystem functions, increasing resilience. Of course, in fire-adapted forests on very dry sites (such as jack and pitch pine barrens), a balance should be struck between leaving deadwood habitats and reducing potential fire risk.

Healthy Soil and Water

Healthy soil and water are the foundation of a forest. Soil with ample minerals and organic matter that is neither compacted nor eroding provides filtered water and critical nutrients to forest plants. A lack of organic matter in the soil can reduce the amount of

Red spruce is predicted to be less competitive in parts of our region due to climate change–driven shifts in its habitats.

available nutrients, the soil's water-holding capacity, and the soil's structure. Compacted soil will increase the amount of stress a tree is under, making it more vulnerable to additional stressors. Soil erosion can reduce the fertility of the soil, limiting tree growth and vigor while degrading water resources. Forested buffers along water resources help maintain water quality by filtering runoff, keeping water temperatures cool, and adding leaves, branches, and logs to the water to be used by a variety of fish and other aquatic species.

Protection for Threatened, Endangered, and At-Risk Species

Species differ in their response to pests and weather disturbances. In general, the higher the species diversity, the higher the resilience of a forest. Species are also connected to and depend on one another. Maintaining our full range of native species across the landscape improves the chances of species and forest adaptation by promoting genetic diversity and the connections between species. This can be done by protecting threatened, endangered, and at-risk species. One way to maintain trees and plants predicted to be more vulnerable in the future would be finding areas of the landscape where they may still be competitive (for example, red spruce in higher-elevation areas in Pennsylvania and Massachusetts) and encouraging their perpetuation there.

FOREST RESILIENCE AT THE LANDSCAPE LEVEL

As we have discussed throughout this book, the context of your property has a profound influence on your forest. The decisions you and your neighbors make about your forests impact your landscape, and considering the characteristics of the wider landscape is essential to understanding your land's resilience.

Landscapes with high resilience are heavily forested and well connected, allowing for the movement of species and genetic material. The rate of conversion from forest to other uses is low, allowing for the persistence of well-connected forest cover. The deer population in resilient landscapes is within a normal range, allowing for timely and sufficient forest regeneration. There are also low levels of other stressors such as invasive plants and insects and disease, allowing forests to function normally. Resilient forests have a diversity of environmental conditions. Some areas within the landscape are mesic (wet) while others are xeric (dry), providing diverse growing conditions that will naturally favor a variety of plants and trees across the landscape. Finally, landscapes with a diversity of forest stages and species composition are well equipped to retain their forest benefits through disturbances.

Conversely, a landscape with low resilience has a relatively small amount of forest and the forests are not connected, making it difficult for species, both plants and animals, to move among the patches of forest and limiting their ability to survive and exchange genetic material. Stressors such as deer herbivory, invasive plants and insects, and disease inhibit forest regeneration and reduce the forest's ability to develop complexity through time. Species diversity is further reduced in landscapes with uniform site conditions.

FIELDWORK

How Resilient Is Your Forest?

Review the characteristics in "Resilience Within a Landscape Context" on the facing page, and then walk your woods. In what ways is your forest resilient, particularly in the context of your goals and values, and what are its vulnerabilities? Where on the resilience gradient do you think your property falls?

Now consider the landscape around your property. How much resilience does the surrounding land have? Does this context change the way you think about your own property's resilience?

Review your estate plan (see Chapter 12) and determine whether it addresses all of your wishes for the future ownership and use of the land.

RESILIENCE WITHIN A LANDSCAPE CONTEXT

High Resilience
LOW VULNERABILITY

Low Resilience
HIGH VULNERABILITY

PROPERTY-LEVEL CHARACTERISTICS

1. Formal plans for continued use
2. Complex forest structure
3. Diverse forest composition
4. High proportion of well-adapted species
5. Healthy soil

SURROUNDING LANDSCAPE

6. Low conversion rates
7. Large continuous areas of connected forests
8. Diverse soil and growing conditions
9. Low deer populations
10. Low invasive plant, insect, and disease pressure

PROPERTY-LEVEL CHARACTERISTICS

1. Future ownership/use is uncertain
2. Simple forest structure
3. Forest dominated by a few species
4. Low proportion of well-adapted species
5. Compacted and/or eroding soil with low organic matter

SURROUNDING LANDSCAPE

6. High conversion rates
7. Fragmented forests
8. Uniform soil and growing conditions
9. High deer populations
10. High invasive plant, insect, and disease pressure

GOALS FOR FOREST RESILIENCE

The general concepts of forest resilience can be categorized into four goals:

Goal 1: Keep forests as forests and connected. Ensure connected, conserved forest cover, preferably on a range of environmentally diverse sites.

Goal 2: Reduce stressors. Limit the amount of stress that forests face to increase forest vigor.

Goal 3: Reduce vulnerability. Make forests more resilient by establishing complex structure and composition and by reducing the potential severity of forest disturbances.

Goal 4: Provide refuge. Maintain the maximum level of plant and animal diversity over time.

Importantly, incorporating these characteristics into your forest still leaves ample room for a broad range of additional landowner goals. In other words, you can improve your forest's resilience and still achieve your unique goals. With an understanding of the characteristics on your property and within your landscape that contribute to its resilience and vulnerability, you can get a sense of where your land falls on the resilience gradient. Remember that your land's resilience is not static; it changes through time. Increasing characteristics that make it resilient will move your land farther left on the gradient. New vulnerabilities to your land or landscape will move your land farther right. The good news is that you need not stand by and let stressors drive your forest's health and well-being. You can increase resilience and reduce vulnerability. Doing so will help you reach your personal goals for the property and help protect the public benefits your forest provides.

Below are the four main goals of forest resilience, each one followed by specific characteristics that can increase the resilience of your forest and actions you can take to restore the characteristics your forest lacks. A consulting forester can help you determine the best strategies for your land.

GOAL 1: Keep Forests as Forests and Connected

Resilient Characteristic: *Conservation-based estate planning has been implemented to ensure the continuation of this land as forest into the future.*

Action: Read Chapter 12, and then engage in conservation-based estate planning to ensure the continuation of the land as forest. Contact a local land trust or estate planning attorney with land conservation experience to investigate your options for keeping some or all of your land in its natural state, including:

- Specifying your wishes for the land in a will or trust
- Changing the ownership of your land (for example, to a trust) to make sure it is passed on according to your wishes
- Donating or selling a conservation easement/restriction on your land to eliminate development on some or all of it

- Donating or selling your land to a conservation organization
- Enrolling your land in your state's current use tax program to help you keep the land undeveloped and reduce property taxes

Resilient Characteristic: *The property is part of a resilient forest or serves as a connection between large areas of forest (more than 250 acres in landscapes with significant development, more than 500 acres in forest-dominated landscapes).*
Action: Conserve resilient forests and the connections between them. Determine whether your property is part of a resilient forest or serves as a connection between large areas of forest by working with a forester, a land trust, an open space committee, or a conservation commission. You can also refer to an online resource to determine your property's landscape position and role.

See your land as part of a bigger landscape. Conserving large, intact, and resilient forests, as well as the connections between them, helps plant and animal species move to habitats that best fit their needs. These areas are also most likely to recover from extreme events, including droughts, ice storms, windstorms, insect outbreaks, and flooding.

Work with your neighbors and local land trusts to encourage conservation of these large areas of land and the connections between them. Examples include:

- Inviting your neighbors to walk your land
- Hosting a neighborhood meeting to talk about the things you like about your neighborhood forests and ways to conserve them
- Reviewing maps and other information to better understand your land and the surrounding landscape
- Connecting your neighbors with resources and local contacts so that they know their options for conserving their land

GOAL 2: Reduce Stressors

Resilient Characteristic: *Invasive plants are* not *found on or near the property.*
Action: Prevent the introduction of invasive plants, remove small populations of existing plants, and learn to manage extensive areas of infestation.

Begin by learning how to identify the most harmful invasive plants in your area, then look for them when you are out on your land. It is far cheaper, easier, and more effective to control a few individual plants than to try to treat a larger invasion. Here are additional tips for keeping invasive plants out of your forest.

- Clean clothing, boots, tools, and equipment that may carry seeds or fragments of invasive plants.
- Work with neighbors to control invasive plants, since controlling them on your land will do little if they are not controlled across the stone wall. Help your neighbors identify invasive plants, and share how you are working to control them.
- Before implementing forest management, take an inventory of invasive plants, and plan for their control and management

before you harvest trees. Allowing sunlight into a harvested forest will trigger invasive plant growth and encourage its spread.

- When conducting a timber harvest, require the timber harvester to power-wash machines before entering your woods.

Resilient Characteristic: *Invasive insects and tree diseases are* not *found on or near the property.*
Action: Prevent the introduction of invasive insects and diseases, and limit the impact of existing ones.

Learn about damaging invasive insect species and diseases in the region, then monitor your woods and community. Report any sign of these pests and diseases to your state forestry agency (see Resources, page 264). Here are some other ways to limit the spread of invasive insects.

- Increase the number of tree species that aren't hosts to these insects.
- When camping, always buy your firewood from local sources rather than bringing it with you. Many pests are unknowingly spread when firewood is moved.

Resilient Characteristic: *There are* no *signs of significant deer impacts or an increasing deer population.*
Action: Work to limit the effects of deer browsing and increase forest regeneration.

- Host deer hunting on your land to control deer populations. Work with your neighbors to implement a hunt across your landscape.
- Leave whole any treetops that have fallen to the ground or have been felled as part of a timber harvest, to shelter seedlings from browse.
- Build slash walls or erect temporary fences in the woods around pockets of ecologically and economically important tree seedlings until they grow tall enough to avoid excessive browsing.
- Apply deer repellents or physical barriers, such as plastic tubing or bud caps, to individual seedlings of desired tree species to minimize the effects of browse.

Resilient Characteristic: *Soils and water are healthy. The soils are* not *compacted or exhibiting evidence of significant erosion.*
Action: If motorized recreational use is one of your goals, reduce the effects of ATVs by directing their use to appropriate locations, such as very sandy or rocky sites, and blocking entry points to other more sensitive areas.

When planning a timber harvest, work with a consulting forester to develop a contract with loggers that includes language about minimizing soil disturbance and using forestry best management practices. Requiring a performance bond from the logger will help ensure that the work is done according to the contract or specify recourse in the event it is not. A consulting forester can supervise the timber harvest to make sure that the requirements of the contract are being met. If the site isn't stabilized according to the contract, the performance bond provides the opportunity to pay someone else to properly stabilize the site.

- Plan skid roads and trails to minimize their number and impact.
- Work only during stable ground conditions, such as when soils are dry or frozen.
- Stabilize the site after the job by installing water bars and seeding exposed soil with a cover crop.
- Leave as many tops and limbs as possible to contribute to soil fertility and help stabilize slopes against erosion.

GOAL 3: Reduce Vulnerability

Resilient Characteristic: *The forest has diverse tree species of various sizes, ages, and spatial arrangements.*
Action: Focus on the successful regeneration of multiple species of trees, and work toward structural diversity.

- Establish or maintain at least two age classes of trees by regenerating portions of your forest.
- Create gaps in the canopy to let sunlight reach the forest floor. The canopy gap size will depend on which species you are trying to regenerate. For example, sun-loving early-successional species need large gaps of greater than 0.5 acre, whereas shade-tolerant late-successional species need gaps created by felling individual mature canopy trees.
- Convert plantations to mixed species forests with native tree species that are well adapted to the site.

Resilient Characteristic: *The forest is largely dominated by species predicted to be well adapted to future conditions.*
Action: Promote the establishment of well-adapted species. The following guidelines suggest species to promote within each kind of forest.

- *Beech-Birch-Maple Forest:* On warmer, drier sites, such as south-facing slopes, promote drought-tolerant species compatible with this northern hardwood forest type, such as red oak. On cooler, moister sites, such as northern slopes, continue promoting traditional northern hardwood species, such as sugar maple and yellow birch.
- *Oak-Hickory Forest:* Central hardwood species such as oak and hickory are well adapted to warmer, drier sites. Therefore, promote the central hardwood species that will best meet landowner or land manager goals.
- *Oak-Pine Forest:* This transition hardwood forest type is typically a combination of species from central and northern hardwood forests. Promote species found in this forest type that prefer warmer sites, such as red and white oak, white pine, and black birch.

Forests with a diversity of species and ages, including ample regeneration, will be resilient to a range of disturbances.

- *Spruce-Fir Forest:* Spruce-fir is a very vulnerable forest type in southern portions of our region, like Pennsylvania, New Jersey, and southern New England. If you are working in this region and have spruce-fir forest on northern, cooler slopes, work to reduce stressors such as invasive plants and insects, and consider turning the area into a small reserve without any active forest management, with the goal of maintaining this forest type. If you have spruce-fir forest on southern, warmer sites, promote species comparatively better adapted to the changing climate, such as northern hardwood species.
- *Red and White Pine Forest:* Red and white pine both occur across a range of sites in parts of our region, and they often predominate on drier mesic soil and can tolerate the drier conditions projected in our region. However, these forests often have a higher density of trees than in the past, given suppression of fire. If you have these forests, work to restore more open, complex conditions to reduce resource competition for moisture and minimize fire risk.
- *Jack and Pitch Pine Barrens:* Jack pine and pitch pine predominate on dry, fire-prone sites, like those with sandy soil or shallow soil over bedrock. In many respects, these species are well adapted to future drought and fire, given that they have evolved with these stressors for millennia. Restoring more open woodland

conditions to these sites will help reduce drought stress and allow fires to burn at low severity to reduce negative impacts.

Resilient Characteristic: *The forest contains few preferred host species of invasive insects or diseases threatening the area (such as white ash, host of the emerald ash borer; eastern hemlock, host of the hemlock woolly adelgid; and red and sugar maple, hosts of the Asian long-horned beetle).*
Action: Reduce the number of trees that serve as host species for invasive insects and diseases in a manner that considers the overall health and function of your forest. Remember, however, that removing host species from the canopy without accounting for long-term impacts on regeneration and understory plant communities could cause unwanted changes to your forest. Therefore, their removal should be part of regeneration harvests or thinning treatments designed to encourage the establishment and growth of non-host tree species that meet your long-term objectives.

Resilient Characteristic: *There are* no *areas of the forest with dense, crowded tree stems.*
Action: Thinning your woods will decrease tree competition for sun, water, and growing space and increase the vigor of the trees that remain. Thin areas of the forest that are crowded, leaving trees of good vigor and desirable species that help you meet your overall goals for the land. The intensity of thinning can also be varied across your forest to create a range of spatial conditions and environments. Some forest types, like jack, red, and pitch pine and oak-hickory, often exist on sites that are better suited for supporting more open forests, like woodlands and barrens. Consult a forester to determine if restoration of these low-density conditions makes sense for its underlying ecology.

Resilient Characteristic: *There are five or more large snags (more than 16 inches in diameter) per acre and five or more large downed trees (more than 16 inches in diameter) per acre.*
Action: Increase the amount of deadwood on your property.

- Take a passive approach and allow the forest to accumulate deadwood over time as trees die from storm impacts and insects. Note, however, that it will likely take decades for these natural processes to accumulate a critical amount of deadwood.
- Reduce the amount of time it takes to develop snags and downed wood by taking an active approach: Create standing deadwood by girdling low-quality or low-vigor trees, and create downed deadwood by felling and leaving low-quality trees as large logs on the ground.
- Protect deadwood during a timber harvest. As long as it is safe to do so, leave snags standing and do not remove downed deadwood from the woods.

Resilient Characteristic: *Water resources have forested buffers to protect them.*
Action: Restore and protect riparian zones by allowing land next to rivers, streams, and other water resources to regrow into forests that will provide soil stability and shade to keep streams cool. In addition, forests add twigs, branches, leaves, and logs into streams, providing food and structure for fish and other aquatic species.

If planning a timber harvest, follow the best management practices (BMPs) for streamside areas. Contact your state's bureau of forestry to request a copy of the BMP manual. See Resources (page 264) for contact information.

GOAL 4: Provide Refuge

Resilient Characteristic: *The property is habitat for threatened, endangered, or at-risk species.*
Action: Determine if your land contains any threatened, endangered, or at-risk species, and identify measures that you can take to ensure their survival. Prioritize these areas for land protection, open space planning, and conservation zoning efforts.

Make efforts to maintain trees at risk of being eliminated due to invasive insects and diseases (such as white, green, and black/brown ash) and preserve their genetic pool within the landscape. If you're in an area that has been impacted by emerald ash borer, start by identifying those trees that have survived the insect. If you're in an area not yet impacted by emerald ash borer, be sure to retain some ash in your forest. Create a "preservation patch" of 12 to 15 ash trees over a couple of acres, and consider treating the preservation patches with insecticide. Collect and store seeds so that you'll have the opportunity to regenerate ash in the future.

Sometimes the best way to conserve something is by taking a passive approach. Small reserves are areas of unmanaged forest surrounding important ecological resources. These reserves can be as small as 10 to 15 trees or as large as dozens of acres. Place reserves around rare plants, endangered species habitats, areas of high diversity, and high-value habitat features (for example, vernal pools). Mark reserves in the field

A forester injects an ash tree with an insecticide to protect it from emerald ash borer. This work is part of a coordinated effort to keep preserved patches of mature ash across the landscape.

with tree-marking paint, and note the locations on a map to ensure they are maintained through time.

Resilient Characteristic: *The property can harbor species that we may otherwise lose from the landscape.*
Action: Identify and conserve microclimates in your forest.

Areas with variations in topography, such as mountains, hills, and valleys, create microclimates that are different from the general climate of the area. These microclimates provide the opportunity for plant and animal species to find suitable habitats to meet their needs. Microclimates are most effective at providing a refuge for species when they are part of large, intact forests that are connected to other large areas of forest, allowing plants and animals to move through the landscape. For example, red spruce is predicted to become less competitive in the southern end of its range because it competes better in cooler growing conditions, but northern slopes may provide a microclimate that is cooler than the surrounding area, offering an opportunity for red spruce to persist in the landscape.

- Conserve resilient areas of diverse topography, geology, and local connectivity to provide options to plant and animal species. Refer to mapping tools identifying resilient lands, and work with neighbors, local officials, and land trusts to maximize the conservation of resilient lands.
- Identify tree species and other natural communities at risk of being lost from the landscape that are currently growing in areas well suited to sustaining them. Take action to conserve those species. This may mean designating an area as a reserve in the hopes that it will persist, or promoting its survival through active forest management.

MONITOR AND EVALUATE

Your forest is a dynamic ecosystem that is constantly changing. In addition, the stressors facing your forest will continue to advance and evolve. Monitoring your forest helps detect stressors early, allowing time for remedial actions. Monitoring can also determine whether or not past stewardship practices have achieved the desired goals. Don't assume your actions were successful! Further action may be needed. If new stressors or vulnerabilities arise or past actions have not led to your desired outcomes, revisit the characteristics of forest resilience and actions to achieve them presented in this chapter, and determine your options. Evaluating and monitoring your forest can be done either informally or formally. Walk your woods. Keep an eye out. Notice how it is changing.

You can also have your forest evaluated by a forester. This can be done informally with a walk-through of your forest or formally through an inventory of natural resources, done by collecting data at different inventory plots spread throughout the forest. Periodic inventories will help you monitor and track

Getting out into your woods regularly is an excellent way to monitor the results of past stewardship practices and to detect new challenges when they're easiest to address.

changes—both positive and negative ones—over time. Regular hikes to monitor your forest are also excellent opportunities to engage your friends, neighbors, and community members by inviting them along.

GETTING STARTED

Understanding where your forest falls on the forest resilience gradient is an excellent start. It is the opportunity for you to learn more about and interact with your forest. Reflect on the values you hope your forest will provide in the future and how those might be impacted by future disturbances or changing climate conditions. Don't be surprised or disappointed if your land isn't fully resilient. All properties have characteristics that make them resilient and those that make them vulnerable.

Also, don't get overwhelmed. Everyone has limited time and resources. Don't think of your list of vulnerabilities as a to-do list. Take the pressure off. Implementing even one single action can have a major impact on your land and the landscape around you. Choose one or two vulnerabilities that are within your goals and resources and work to address them.

Conservation-based estate planning to ensure your forest will remain forest after you are gone should be considered a priority. After all, without forests, there can be no resilient forests. Every time you address a vulnerability, you move your forest farther left on the gradient. Be sure to monitor your forest to make sure those vulnerabilities you identified are fully addressed and that new vulnerabilities haven't arisen.

Practicing Ecological Forestry to Increase Forest Resilience

CONTINUITY: Retaining trees through active forest management will help achieve a diversity of tree sizes and ages. Designating patch reserves will create diverse environmental conditions within your forest—the reserve patch may provide deep shade while areas within a regeneration gap will have greater light and resources for new growth. Designating a variety of tree species as legacy trees will help ensure diverse species composition. Formalizing your conservation-based estate plan to ensure your forest will remain forest is foundational to its resilience.

COMPLEXITY AND DIVERSITY: Increasing complexity and diversity is key to forest resilience. Encouraging forests with multiple canopy layers, including adequate levels of regeneration, will reduce the risk that one disturbance will impact all the trees, and will ensure that new trees are poised to grow into the overstory to replace mortality. Multiple canopy layers can be achieved by encouraging diverse species with different regeneration pathways and a range of moisture and temperature tolerances. Maintaining appropriate deadwood levels provides regeneration substrates.

TIMING: Extending the time between regeneration harvests (10 to 20 years or 2 to 3 inches in diameter of tree growth) will help diversify tree size and age by promoting some large and old trees. At the same time, recognizing that the pace of change affecting forests is more rapid than in the past, you may need to engage in stewardship of your land more often to reduce stressors, increasing resilience.

CONTEXT: Resilience should be considered at both the property and landscape scales. A diverse landscape will include a variety of forest stages and species composition and will be less likely to suffer significant impact from a disturbance. Work to promote forest cover in your landscape and retain connections between forest blocks to facilitate the movement of species and genetic material.

CASE STUDY

Increasing Forest Resilience

Janet Sredy

LOCATION: Elizabeth, Pennsylvania

ACRES: 110

Janet Sredy, her brother, and their respective spouses demonstrate one family's efforts to increase forest resilience by securing full ownership, addressing vulnerabilities, and promoting species diversity while also increasing their own enjoyment of their land.

Their 110-acre property has been in the family for 100 years. After Janet's great-grandfather's death, the land was divided among many relatives. Her father had been slowly repurchasing deeds from other family members. By the time Janet and her brother inherited the land, they had five-sevenths of the original property. Yet the remaining sections, now owned by people outside the family, were crucial for accessing their land. By negotiating with the neighboring landowners and starting a legal process, Janet and her brother obtained the last two parts of the original property. Janet and her husband are the primary managers but have formed an LLC ownership with Janet's brother and his wife, who live across the country. With the ownership secured through the LLC, they are now working on formal plans to keep the land intact and forested when it's their turn to pass it on.

Although piecing the original property back together was a success, the land was in rough shape when it was passed on to them. The forest was so tangled with invasive plants, such as Asian bittersweet, they were unable to walk through it. The loss of ash trees to emerald ash borer led to the establishment of multiflora rose in the gaps where the ash had been. As a new landowner, Janet did not know where to begin.

She started by searching the internet and found Pennsylvania's Natural Resources Conservation Service (NRCS) and the Department of Conservation and Natural Resources. They encouraged Janet to contact her local state service forester, and she arranged for him to visit and walk the property. The

> Treating the invasives led to the property being designated as a wild plant sanctuary.

forester encouraged Janet and her brother to get a management plan so that they could outline their goals and develop steps to achieve them. At first, they hesitated, but they ultimately decided to hire a consulting forester and move forward. They have been on a roll ever since.

The consulting forester identified a local contractor with the necessary equipment to treat the invasive plants. He used a skidder-mounted sprayer to treat them and returned the following year to spot-treat the forest. Funding for the invasive control included money from the Environmental Quality Incentives Program (EQIP). With the wall of invasives gone, they also were able to identify rare plants, which led to the property being designated as a wild plant sanctuary. The absence of invasive plants also created opportunities for them to plant a diversity of tree species to help occupy the vacated space and discourage future establishment of invasives. The lack of invasives also allowed them to move forward with wildlife habitat work, the establishment of a trail network, and timber harvesting.

Clearing invasives along the property lines has enabled them to both reestablish clear boundaries and meet their neighbors. The resulting conversations have led to them helping their neighbors with forest stewardship, including tree planting and invasive plant control. Janet and her husband plan on spending more time with their neighbors, helping them steward their land and advancing their mutual goals.

Amid all this effort, Janet and her husband enjoy the solitude, the wildlife, and the challenge of the work itself. Seeing the improvement of the land makes it all worthwhile, but going forward, they will try to take their own advice and sit back and enjoy the property even more.

"They live on the land, but not by the land."
—Aldo Leopold, *A Sand County Almanac*

CHAPTER 10

Local Wood, Local Good

IF YOUR INCLINATION IS TO SKIP THIS CHAPTER because wood production is not one of your goals, please reconsider. Landowners often profess to want to do the "right" thing for their woods, and the idea of harvesting trees and producing wood products can sound contradictory to doing the right thing. That's what makes this chapter important.

While there is broad recognition that humans need forests for many life-sustaining benefits, including water quality and climate change mitigation, there is less acknowledgement that humans need wood products. Nevertheless, the fact remains that we depend on products made from wood for much of what we do in our daily lives. So, how do we go about balancing the human need for wood products and their harvest from the forest with a commitment to honoring and stewarding forests in a way that helps restore them and their function?

Producing and using local wood is an opportunity to build community.

FINDING A NEW BALANCE

The first step is to acknowledge that our consumption of wood products has both ecological and social impacts at local, national, and global scales. We must take responsibility and seek to meet our need for wood in a way that mitigates these impacts. To do so will require a cultural shift in our society, a shift that must be led by forest landowners.

To be clear, this isn't a defense of all timber harvests. Exploitive timber harvesting certainly does exist, and it seeks to remove only financial value from the forest with little to no effort directed at improving the forest. However, ecological forestry provides guidelines for balancing our need for wood with our duty to enter into a reciprocal relationship with forests, helping to restore and sustain them. The principles of ecological forestry—continuity, complexity and diversity, timing, and context—provide a framework to keep the extraction of resources within forests' natural limits and to work toward reciprocity. Our forests can provide the wood we need, if we harvest it locally and ethically.

While we hope that this chapter will inspire more landowners to engage with their forest through active forest management, it is not meant as a condemnation of the passive approach. As discussed throughout this book, the passive approach is not in opposition to the active approach but is complementary to it. They are parts of a larger whole, and we need both. You get to decide the best approach for your goals.

There are compelling reasons to meet our own wood needs locally and ethically. As a landowner, you need to decide: How will you balance the reality of your personal wood consumption and that of our society with your opportunity to contribute to the production of wood products? In what ways might your forest management practices reciprocate for the wood products produced by the forest? How can harvesting trees for wood products wood products result in more than just logs (e.g., wildlife habitat, increased resilience, connection with neighbors)?

THE ILLUSION OF PRESERVATION

While the harvest of locally grown produce is celebrated by the public, as it should be, the harvesting of local trees is often seen as the exploitation of forests for economic gain. Those without an understanding of forests and forestry assume that, like the Onceler in Dr. Seuss's book *The Lorax*, anyone who is cutting trees is making money while degrading wildlife habitat and water quality. Farmers are also given respect for their role in meeting the needs of society, again as it should be, while foresters and loggers often feel unappreciated or even vilified for cutting down trees. Landowners themselves may also come under fire from neighbors and others in the community for implementing a timber harvest. Those who support timber harvesting are rarely seen as environmentalists. With such negative perceptions of timber harvesting, it must be safe to assume that those opposed to timber harvesting don't use wood in their daily lives, right? Wrong.

As a nation, we are voracious consumers of wood products. According to US census data, in 1975 the average new single-family house in the northeastern United States was about 1,600 square feet; today it's nearly 2,300 square feet, even though the average family size has decreased. In addition, the convenience of online shopping has led to increases in packaging for all of those items that magically appear on our doorsteps. In fact, our consumption of wood exceeds the consumption rates of other countries with similar economies, such as Japan, Germany, and Finland. Likewise, our production rates lag behind these same countries. In short, we love to use wood, but we don't like to produce it, or, maybe more accurately, we don't want to see it being produced in our backyards.

The "illusion of preservation" is the false assumption that eliminating active forest management from local forests leads to the preservation of the environment. However, without a corresponding reduction of wood consumption, efforts to stop timber harvesting only result in the relocation of timber harvesting from the locale in which wood is being used to other parts of the country and the world. Out of sight, out of mind. Rather than taking responsibility for the sustainable production of our own resource needs, when we stop timber harvesting we lay that burden on others.

Impacts of the Illusion

There are significant ecological, social, and ethical consequences to the illusion of preservation. Pushing timber harvesting to regions with less environmental oversight and limited requirements for best management practices is likely to have impacts on soil health and water quality and the local communities in those areas. Biodiversity can also be compromised through degradation of soil and water and a lack of cutting standards that protect important habitats and species. Transporting wood long distances from where it is grown to where it is being used increases the carbon impact of the materials. Additionally, the carbon cycle is a global system. Not harvesting trees in one region to maintain forest carbon storage while forcing timber harvesting to move to another region to supply the wood needs of those not harvesting is called leakage. The carbon benefit is an illusion.

Forest management is an important part of the economy of some rural communities and can help maintain the traditions and relationships we have with the land. The shift of responsibility for our consumption may include impacts to Indigenous peoples both at home and abroad. There is also an elitist attitude embedded in the illusion of preservation that should be called out. Those with affluence often have high consumption rates but can afford to push production onto those with fewer economic advantages. The ecological and social impacts of the illusion provide a strong moral imperative that we produce the wood we use.

Dispelling the illusion of preservation necessitates achieving two primary goals. First, we need to reduce our overall wood consumption. Our consumption of wood puts additional pressure on our forests. Lower rates of consumption will reduce the production pressure, helping us move from what is predominantly a timber-based management approach in many regions to the implementation of ecological forestry. Second, we must embrace local production. Humans need wood products. There is no way around this fact. The most ecologically and socially responsible way to meet our wood product needs is to do so as locally as possible using management strategies that sustain forests. Achieving these goals requires examining perceptions about timber harvesting and promoting the widespread adoption of ecological forestry among landowners. As a landowner, you play perhaps the most critical role in dispelling the illusion of preservation.

VICTORY GARDENS FOR FAMILY AND COUNTRY

An interesting model for local forest management could be America's wartime victory gardens. World War II posed an existential threat to our country. There was great concern that the war would destroy the world, and even if the world survived, an Axis victory would lead to the loss of our way of life. The shocking attack on Pearl Harbor brought the reality of a distant war to our home soil and led to the US entry into the war. Our entire country mobilized to meet the challenge. The soldiers on the front lines met the enemy head-on, many of them making the ultimate sacrifice. On the home front, everyone was expected to do their part for the war effort. Those who lived through WWII are called the "greatest generation" in recognition of their determination and sacrifice and their faith that they could overcome the daunting challenges brought on by the war.

One of the efforts implemented on the home front was the victory garden. A concept that originated in WWI and was dubbed "victory garden" in WWII, its goal was to compel people to grow and preserve food, helping to stretch food supplies for both those on the war front and those at home, all of whom were on rations. Homeowners and landowners were encouraged to increase the amount of food they produced. Community gardens were started in urban areas. The US government's goals for victory gardens are illuminating and offer insights into our current challenges:

1. Increase the production and consumption of fresh vegetables and fruits by more and better home, school, and community gardens, to the end that we become a stronger and healthier Nation.
2. Encourage the proper storage and preservation of the surplus from such gardens for distribution and use by families producing it, local school lunches, welfare agencies, and for local emergency food needs.

3. Enable families and institutions to save on the cost of vegetables and apply this saving to other necessary foods which must be purchased.
4. Provide through the medium of community gardens, an opportunity for gardening by urban dwellers and others who lack suitable home garden facilities.
5. Maintain and improve the morale and spiritual well-being of the individual, family, and Nation. The beautification of the home and community by gardening provides healthful physical exercise, recreation, [and] definite release from war stress and strain.

The Success of Victory Gardens

So, how did Americans respond to the existential threat of WWII? In 1939, more than $200 million worth of vegetables were grown in 4.8 million home gardens. In 1944, 18.5 million gardeners took part in victory gardens, supplying 40 percent of the nation's fresh vegetables. By the time the war was over in 1945, American victory gardeners had grown between 8 and 10 million tons of food. The success of victory gardens is undeniable.

While the victory garden's sense of urgency and duty has waned over succeeding generations, it hasn't completely disappeared. Today there is a renewed interest in local agriculture. Farm-to-table restaurants have become popular in cities across the country, featuring local agricultural products and menus focused on seasonally grown foods.

Government promotion of victory gardens encouraged people to produce their own food locally.

Beyond meeting the immediate need of supplying food, there is a series of positive ripple effects from growing local food. Growing your own food allows you to do so in a way that matches your own beliefs and ethics. Don't like the use of chemical fertilizers and pesticides? You can grow your food organically, start your own compost pile, and adopt companion planting practices. If you're worried about the excessive use of water in large-scale agriculture, you can install drip irrigation and add organic matter to the soil. The same is true for commercial food production. Not happy about the amount of added sugar and preservatives and the use of dyes? Process and store food in the way you want. Do you want to minimize the carbon footprint of food? Reduce the distance your food travels down from the number of miles it takes to ship it from California to the number of steps it takes you to walk to the kitchen from your garden.

Growing food connects us with people in ways we may not see. Building and maintaining growing space takes materials and tools, requiring interactions with local suppliers. Addressing challenges such as insects and diseases is often helped by connecting with fellow gardeners to learn about the strategies they employ to deal with these challenges. Growing your own food often leads to gifting some of your hard-earned produce to family, friends, and neighbors. There are few greater pleasures in life than sharing a meal made with food you have grown yourself. In short, growing one's own food increases local control and local benefit.

The COVID-19 pandemic increased the trend toward local agriculture and self-sufficiency. It provided a sobering reminder of the fragility of our supply chains and our dependence on them. It would only be a matter of days before supplies got low in local grocery stores if food transportation came to a halt. A local, reliable food supply increases resilience in the face of future pandemics or other societal disruptions, such as a natural disaster, contributing to public well-being and security.

Victory gardens demonstrated what can be accomplished by a motivated and unified citizenry, and it is astounding. A success of this scale, although frankly hard to imagine today, was achieved because Americans' backs were against the wall and everyone pulled together. Reaching victory garden levels of forest engagement to face our current threats is possible. Although the quality of the outreach and educational materials produced to promote and support victory gardens was impressive, it wasn't these materials that drove adoption. Instead, it was a broad societal recognition that we needed to support our country and the world during that time of great uncertainty. The motivation caused by World War II, in effect, caused a cultural shift, a changed perspective among the citizens of the United States.

CREATE YOUR VICTORY FOREST

Think of your forest as a victory forest, with all of the same benefits that come with local food production. Imagine if, using victory gardens as a model, victory forests were established through the application of ecological forestry by millions of landowners across our region, not only in our rural landscapes but also along the rural-to-urban gradient. Every woodlot or patch of trees could be a victory forest and part of the solution, no matter its size or location.

Remember, there are no cookie-cutter approaches in ecological forestry.

Wood for this timber frame was grown and processed locally, and family and friends worked together to raise the frame. The result is not only a beautiful and environmentally friendly home but also people connecting with one another and the local forest.

Landowners would apply the strategies that match the unique qualities of their forest and meet their specific goals. The diversity of approaches taken by landowners would naturally lead to diverse forest stages and species composition across the landscape, resulting in diverse wildlife habitats, resilient carbon sinks, and, as a by-product of these efforts, local wood to meet local consumption. There is a knowledge of the forest that can be gained only through interaction and active engagement. Like all relationships, it takes time and effort to work out the right balance. Also, seeing the effort of wood production and understanding the limits of forests has the potential to bring our consumption into more realistic bounds, as we begin to increase our respect for the gift of wood and honor it with a reciprocal care for our forests.

Building Relationships and Economies

The use of local wood for local good builds relationships when people begin reconnecting with the role forests play in our lives. Landowners connect with one another to share their knowledge and experience. These relationships lead to increased opportunities for landscape management and collaboration, including efforts to control invasive plants and deer herds, implement joint timber harvests, create blocks of habitat and connectivity, and build connecting trails. When wood from the local woodshed is used for the flooring in a community building, a timber-framed outdoor classroom for a school, or even firewood for one's home, it creates an opportunity to connect people with the forests and those who own them, steward them, and create sustainable products from them.

Just as produce must be stored and preserved appropriately to be useful over time, wood must also be processed into a usable form. This necessitates both the production of logs and the infrastructure to process them into ecologically friendly products, which can then be used to make secondary wood products such as furniture, cabinets, and flooring that meet the needs of local communities. All of this labor and manufacturing fuels economic development and innovation.

The local production of wood products can strengthen our regional supply chain, providing resilience in the face of future disruptions. Local wood production even increases national security through decreased

These milled boards have been stacked back together to resemble the log from which they were originally sawn. They are a reminder of the source of the wood products on which we depend and our role in producing them.

dependence on other countries for our wood needs. Finally, by deepening our connection to nature and community, victory forests, like victory gardens before them, can facilitate a shift in society, increasing our mental, physical, and even spiritual well-being through greater connection to forests and one another.

It Will Take a Village

Though family forest owners will need to be the main drivers of this cultural shift, there are important opportunities for all types of landowners to contribute to it, including holders of public lands (federal, state, county, town) and land trusts. These lands are often larger than family forest land and therefore play a unique role in the landscape, including helping to create connectivity among forests and providing demonstration areas. In the same way that school gardens are important, this cultural shift would benefit greatly from the inclusion of young people who can learn the importance of their local forests and the pride of stewarding land for the benefit of all.

Are victory forests a pipe dream? Victory gardens demonstrate that the dream is possible, if we have the will. As a forest landowner, you play a unique and crucial role in helping to create a shift in our culture to one of forest connection and appreciation. Depending on your goals, engagement with your forest may include timber harvesting to create desirable stand structure and species. You are producing necessary wood products while creating your desired condition. Be proud of your contributions to society while producing wood. You are taking on responsibility for our consumption and doing it, using ecologically and ethically sound strategies, in a way that benefits forests and all those who depend on them. Be sure to remind people who are talking about the importance of farm-to-table efforts that the table of which they speak is made of wood.

MORE THAN WOOD PRODUCTS

Meeting regional wood needs with local production is more than simply having local timber harvests. Local timber harvests must be done in a way that sustains forests and people. That is to say, if one is focused only on removing valuable wood products from a forest with little or no consideration to the future of the forest, then although the timber harvest does produce local wood products, it's likely an unsustainable practice that will have deleterious effects on both forests and people. Similarly, if we solely emphasize the role of forests as places that supply wood, even if it is harvested sustainably, we are no longer honoring our relationship with these amazing ecosystems and the many gifts they provide. Pairing the practice of ecological forestry with local production can meet our wood consumption needs and help honor this relationship by addressing the threats of climate change and the biodiversity crisis.

Growing Financial and Ecological Value

As with goals for old-growth characteristics, wildlife, climate change, and resilience, there are specific strategies you can employ to grow high-value forest products with long-lived carbon storage. There is also a broad spectrum of actions that a landowner can take to be a responsible steward of their financial assets. Investing in precommercial thinning—which focuses on improving the growth of desirable tree saplings by cutting competing saplings with no economic value—can help ensure that species that make wood products are well established and free to grow. While these practices can require an investment of time and money, their long-term benefits are great. Commercial thinning that focuses on removing trees of poor quality and low economic value helps improve growing space for the best individuals of the commercial species while also meeting ecological goals tied to restoring large tree structures and trees with abundant carbon stores.

The goal of a thoughtful timber harvest is to create conditions in your forest that will lead to the species composition and structure that will provide the forest benefits you seek. The trees that are removed and the forest products they produce are by-products of creating the desirable forest conditions.

Generally speaking, when conducting a timber harvest, concentrating on removal of poor-quality, low-vigor trees will improve the quality of the forest over time. A timber harvest shouldn't cost you money, since the logs that are being removed have value. Some of the logs will have high financial value. Others will have low financial value. It is very easy to sell large, high-quality trees of commercial species because of their high financial value. It is more difficult to sell trees of low financial value, such as small ones and those with defects. Typically, some higher-quality trees will need to be included in a sale of low-quality trees to make it a financially viable harvest for a logger.

More important than the immediate financial value of what is removed during active forest management is the resulting condition that is created within your forest. The general goal is to remove as many trees of low financial value as possible and only as many higher-quality trees as necessary. Over time, this strategy will increase the financial value of the forest while enhancing structural complexity, species diversity, and forest resilience.

In turn, increasing the financial value of your forest increases your management

CURRENT USE PROGRAMS

Though we have focused heavily on the ecological and social values of your forest, the reality is that your forest is also a financial asset, bringing with it costs and opportunities. In fact, for many landowners, their land is their largest financial asset. Despite the perception that harvesting trees is only about money, protecting and growing the long-term viability of this asset isn't greedy; it's prudent decision-making. Unlike a farmer selling agricultural products each year, very few forest landowners earn income from their forest on an annual basis. On the contrary, instead of receiving revenue from all the public benefits provided, landowners are required to pay property taxes.

Generating periodic income from your forest can offset the carrying costs of land and can supplement your household's regular income. However, in many areas of our region, the amount of money that can be made through the long-term production of wood products is less than the property taxes paid on the property. Land is typically taxed ad valorem or "according to its value," and frequently this is calculated based on its "highest and best use." This results in forestland being taxed as if it were developed for residential use, despite a landowner's intention to keep the land in its natural state. To make the property financially viable, it can be very helpful to combine your forest management with your state's current use program.

Current use programs are a "quid pro quo," or "something for something," agreement. Your forest provides tremendous public benefits, including clean water and air, wildlife and biodiversity habitat, and climate change mitigation. Current use is the general term used to describe a program that reduces a landowner's property taxes in exchange for a commitment to continue providing these and other public benefits. These programs are also known as preferential tax programs, and they have specific names in each state.

Each state's current use program has its own requirements for enrollment and benefits in the form of property tax reductions. They may also have penalty taxes for leaving the program. Many current use programs are implemented by a state agency, while others are implemented at the county or town level. Some states have one program for all landowners, while others have separate programs for agricultural land and forest land.

It is common for forestry current use programs to require active forest management for the production of forest products. They will often require a formal forest management plan and the implementation of timber harvests as recommended by the plan, helping to ensure the provision of public good. These programs can make a significant difference in making the land affordable while helping you learn more about your forest and developing a relationship with a qualified forester. For more information, contact your state's bureau of forestry.

options by providing valuable trees to mix into your harvests to make sure they're saleable. The financial value of your forest can also be used to secure improvements to the forest and its infrastructure—for example, by installing a gate, grading the skid roads of the timber harvest to allow access by a pickup or for recreation, or treating invasive plants. Importantly, trees that are of low value economically can still be of high value ecologically, so cutting and leaving low-value trees as a source of downed deadwood could be one option if your goals and local markets don't support their removal.

Implementing active forest management to improve wood quality provides more than just financial benefits. Many tree species that make valuable wood products also have excellent wildlife value. For example, red and white oak and black cherry are all financially valuable, and their mast is very valuable for wildlife. Similarly, valuable white pine provides a unique canopy structure for many raptor species. Promoting high-quality trees also means increasing the growing space for crop trees to allow for the development of large, full crowns to increase growth rates of the trees. Larger crowns are able to produce more mast and provide an increased area for insect foraging.

The Timber Harvest as a Tool

Producing local wood can and should be about more than simply harvesting trees from the forest. Timber harvesting should be the implementation of thoughtful active forest management strategies to create a desired future condition that will meet your goals. As we have talked about throughout this book, the type and number of benefits derived from a forest are a function of its species composition and structure. The removal of trees is an opportunity to increase light levels to promote residual trees (thinning) and/or to favor the establishment of desirable species (regeneration). Timber harvesting is also a tool to shift forest structure to enhance forest benefits.

Practicing ecological forestry goes beyond standard forestry to ensure that the whole forest is considered through the application of its principles. By using timber harvesting to create a desirable condition within the forest, wood products become the by-product of creating the condition as opposed to the end in itself. This is a 180-degree turn from the exploitive approach of simply removing trees with the highest value or from emphasizing solely the timber-producing trees in a complex ecosystem. For example, if the goal is to regenerate gaps in the canopy to let in enough light to favor a shade mid-tolerant species like red oak, then the focus of the timber harvest should be on that restorative goal. Trees that are removed to create the gap are the by-product. Focusing on the desired condition for your forest will produce wood products.

Incorporating the production of wood products into other landowner and societal goals creates multiple benefits. Rather than simply choosing an area of your forest to harvest because it has big trees, consider opportunities to use active forest management to achieve goals. Are there stands in your forest in which a timber harvest will help you

A forwarder machine removes white pine logs from a gap intended to create the light conditions necessary to establish regeneration of desirable species.

diversify wildlife habitat? Would an active timber harvest help you diversify species composition or structure to increase the resilience of your forest? Are there opportunities to use active forest management to promote old-growth characteristics? If so, these areas would be good candidates on which to focus a timber harvest. Remember to also evaluate the landscape in which your land lies and the role your land plays within it.

DETERMINING WOOD PRODUCT CLASSES

We've been using the term "wood products" in a very general sense. Your trees may be used for a variety of different forest products, depending on their size, species, and quality. Each tree has its own unique wood qualities. Having a basic understanding of the various wood products that can be produced from trees cut from your forest can help demystify the process of a timber sale and give you a sense of the ways in which you are contributing to our society's need. Feel proud that you are providing wood products. It is no less important than providing food. To start, wood products are broken into various product classes. The determination of which trees are assigned to which product class is based on the tree's size, species, and quality.

Log Diameter

Some product classes have minimum diameter at breast height (DBH) limits. Trees below the minimum diameter cannot be considered for that product class. For example, trees typically need to be at least 12 inches DBH to be considered for sawtimber logs and 16 inches DBH

PRODUCT CLASSES AND VALUES

Veneer is the most valuable product class. Veneer logs are peeled into very thin slices of beautifully grained wood and glued onto the face of products made of other, less expensive species. High-quality individuals of certain species of hardwoods, such as sugar maple, red and white oak, black cherry, and black walnut, are all examples of species that can make veneer logs if the size and quality of the logs meet veneer standards.

Sawlogs represent the next most valuable product class. Wood sawn from sawlogs is often used in furniture and other high-value, long-lived products. Within the sawlog product class there are several grades of quality, with grade 1 being the highest. Sawlogs are sold by the board foot. In well managed forests in the mature and old forest stages, the proportion of sawlogs can account for around two-thirds of the total tree volume.

Firewood, like other low-value products, doesn't require trees of a certain size or quality. Certainly, there are species differences—white ash seasons fast, red oak splits well, and hickory is high in Btu—but a wide range of species can be used. They need only to be processed to their desired length to fit in a wood stove. Seasoned firewood can be sold for a higher price than green, or unseasoned, firewood.

Pulp and biomass chips don't need to be derived from high-quality trees. Pulp is wood that has been shredded and broken down by physical and chemical means for use in products such as paper and cardboard. Biomass chips are burned for the generation of heat and electricity. Pulp and chips are often sold by the ton. Markets for these low-grade materials are regionally specific.

Logs are piled on a landing and sorted by product class.

to be considered for veneer. Harvesting a tree before the minimum diameter is a potential loss in financial value, similar to an early withdrawal penalty. There is also an ecological loss, since large-diameter trees store a disproportionate amount of carbon and also provide important habitat lacking from most forests.

Log Length

A tree's diameter is not the only consideration when evaluating its size. Some product classes include a minimum log length. Once the tree is felled, it will be cut into shorter sections. This is called bucking the log. The logger will make bucking decisions based on the length of logs needed for the product class in general and, more specifically, the length needed by the mill to fill an order. For example, a mill may need to fill an order of white pine boards that are 12 feet long. The logger will then do their best to identify as many 12-foot logs in the harvest as possible. It is also customary for loggers to buck logs slightly longer to account for any trim on the boards that need to be made.

Of course, it's not possible to turn every tree into 12-foot logs. Logs of other lengths will be used for other orders. The bucking decisions made by the logger have a financial impact on the timber sale, since maximizing the logs for their highest-value use will result in a greater return, particularly in high-value logs.

Tree Quality

Just because a tree is of a certain size doesn't mean that it will automatically qualify for a product class. The quality of the tree is another important consideration. Tree quality is determined by the amount of rot in the tree, its straightness, and the number of clear faces, or sides, without branches, knots, or other blemishes. There are very specific grading standards to determine the number of clear faces when considering trees in the woods. The more clear faces on the log, the higher the product class it can qualify for.

The quality of a tree can be influenced by pruning, another precommercial activity that requires an investment of time and/or money but pays off in the long run through higher quality logs. Removing the lower branches when the tree is young will help the tree develop clear faces on the log. The first log (16 feet) of a tree is typically the most financially valuable, so pruning is customarily done on the first 17 feet, allowing for an extra foot, since trees are not cut flush with the ground.

Species

The tree species also helps determine the product class. Each tree species has its own characteristics. Trees that grow fast, such as shade-intolerant pioneer species, tend to have weaker wood, and those that grow slowly, like shade-tolerant species, typically have stronger wood. Depending on the specific wood use, the product may need a certain strength. For example, if the wood will be used for building, it needs to be structurally sound to bear weight. Different tree species also have different color and grain patterns. These aesthetic considerations change with styles over time, as can be seen in preferences for products such as kitchen cabinets and flooring.

Logs are processed into primary products through sawing, edging, trimming, and sorting boards.

Sometimes lighter wood, such as sugar maple or paper birch, is in fashion, and other times darker wood, such as black cherry or red oak, is preferred. These shifts in trends can affect the demand for species and the price people are willing to pay for them.

Based on this combination of size, quality, and species, a tree is sorted into a product class—typically veneer (the most valuable), sawlogs, firewood, pulp, or biomass. The goal is to use the log from each tree for its highest and best commercial use. Doing this will not only maximize the financial return, but it will also move logs into product classes where the products have some longevity and therefore will store carbon for the longest period of time possible, keeping it out of the atmosphere. In addition, this focus on highest and best use serves as a way to fully honor the contributions that a tree can make to our lives as a forest product.

Achieving a Desired Future

While trees in the lower product classes have less value, the removal of these trees can be a very important component in creating the desired future condition within a forest. In fact, it is not uncommon for the low-value product classes of firewood, pulp, and biomass to be sold for very little or even no money, as their value isn't in the product itself but in the condition it creates in your woods, such as increased growing space for better-quality trees or gaps in your forest to regenerate more desirable species. In many timber sales, the logger's cost to remove these low-grade trees means you at best break even. Your consulting forester can advise you on the current value of these and all of your wood products.

It would be a financial loss—and a dishonor to the legacy of the tree—to chip a tree that is large and high-quality enough to be a sawlog and use it for pulp. A good logger will work hard to find the highest and best market for each log. In doing so, not only is their return maximized, but often so is yours. Loggers will put logs into specific piles on the land based on the products that the logs will be sold as; this is called their sort. Loggers with a number of different piles, or a finer sort, are often maximizing the value of each log.

Using forest management to shift the proportion of trees in your forest toward more valuable species and product classes does more than increase the financial value of the trees. It also means that more of the trees in your forest will store carbon for longer. For example, firewood will do a valuable service as a locally sourced fuel, avoiding the impacts of heating oil, electricity, natural gas, or propane, but its carbon storage is limited to the time between its harvest and its burning. In contrast, veneer trees and sawtimber are typically used for buildings and furniture, and carbon will be stored in their wood for decades, if not longer.

GROWING THE BIOECONOMY

Wood is a very ecologically friendly product. There are efforts around the world to expand the uses of wood. These efforts are a subset of the economy and referred to as the bioeconomy. The bioeconomy is a fast-growing sector that focuses on renewable, biologically derived goods and services to create sustainable business models that are environmentally friendly. Wood-based examples include cross-laminated timbers and other engineered wood products that can replace carbon-intensive materials like steel beams in

Stacked boards are stickered with wedges to allow air to flow around them and help the boards dry evenly.

construction. Other examples include wood fiber insulation and wood-based substitutes for molded plastic, including plastic bottles and food packaging.

Producing environmentally friendly products, especially those that can substitute for environmentally damaging products like plastic, is a benefit to us all. By expanding our reliance on forests, the bioeconomy offers opportunities to increase the connection between people and forests by further demonstrating the essential role forests play in meeting the needs of our society.

More specifically, the bioeconomy offers particular opportunities to landowners. Beyond the traditional wood products described above, it is likely that the future will include markets for wood products that we cannot even conceive of at present. Selling high-quality, large trees will never be a problem. The more markets there are for trees of low financial value, the greater the opportunity there will be to implement forest management strategies to create desired conditions in the woods and break even or even make money while doing it. The greater society's reliance on forests, the greater the likelihood of public policies that support landowners in producing these products.

FIELDWORK

Can You Produce Wood from Your Forest?

Take a walk in your woods. Which parts of your forest are your favorites? What do you want to know more about? What resources or people might help you better understand those topics? Determining other goals you might have for your forest is a good first step to finding out how those goals could be joined with the goal of producing wood products.

Start small: Do you have interest in using wood from your own forest? Perhaps you are interested in harvesting your own firewood. Or maybe you have a building project in mind and would like to use wood from your own property. Consider the possibilities and talk to your consulting forester about how active management can help you reach your goals.

GETTING STARTED

For many landowners, the jump from low engagement to a thoughtful timber harvest is often daunting. That's very understandable. It would be like jumping from a first date to the wedding. As described above, your forest management should not be primarily focused on cutting trees to produce wood products but on creating a desired forest condition to meet your goals. So, once again, take the pressure off yourself. The cultural shift is first and foremost about your relationship with the forest, so first get to know one another. Start by taking just one step toward greater engagement with your forest and go from there. Actively pursue your interests, be it invasive plants, foraging mushrooms, trails, or wildlife. Allow your questions and interests to guide you. Often the first step will lead to obvious and natural next steps.

Practicing Ecological Forestry for Local Wood Production

CONTINUITY: Focus more on what you're leaving behind after a timber harvest than on what you're removing. Leaving biological legacies and a healthy amount of deadwood provides for the next generation of trees by protecting existing regeneration and providing a variety of environmental conditions for new regeneration.

COMPLEXITY AND DIVERSITY: Rather than simplifying our forests to favor a small number of species of current interest to wood markets, promote a diversity of species to allow for a diversity of possible forest products, including the development of unforeseen wood products within the bioeconomy.

TIMING: Extending the time between regeneration harvests not only increases in-forest carbon storage and gives habitat a chance to develop, it also grows larger trees that have a greater likelihood of raising the value of your forest products. Of course, there is some amount of financial risk to leaving the trees in your forest longer, since they could be lost to a disturbance. As a landowner you will have to balance your goals with forest reciprocity. High-value forest products have greater potential for long-term carbon storage in durable wood products. These higher-value products also provide more opportunities to add value in support of local economies.

CONTEXT: While a diversity of age classes and forest types across the landscape promotes wildlife and biodiversity, resilience, and forest carbon, it also helps to provide a flow of wood products over time. A diversity of forest stages and species across the landscape will also help reduce the likelihood that one large disturbance will result in the widespread lack of wood products.

CASE STUDY

Sharing Lumber . . . and a Meal

Ed Neuhauser

LOCATION: Groton, New York

ACREAGE: 132

Growing up near New York City, Ed Neuhauser knew from a very young age that city life was not for him. He yearned for the woods, and he has made that desire a reality. He and his wife now own 132 acres of forest in central New York, a world apart from city life. The land gives Ed the chance to learn about the forest and opportunities to steward it and experiment with different forest management strategies, all while producing necessary products from it to help friends and family. Ed enjoys not only implementing his own forest management but also milling the logs from his forest on his bandsaw mill. Managing the forest and producing wood products is an opportunity to connect with his forest and people.

Ed has a passion for working with wood. The wood resonates with him. It calls to him. A trained scientist, he enjoys the puzzle each log presents as he works to find the best way to saw it up to get the largest possible yield from the log for the highest-quality project. The boards and beams that come from the logs on his property have been used to build outbuildings for his own home as well as for projects across a much larger geographic region. Ed's woods have provided flooring from species such as black cherry, American elm, and even American beech. He has given flooring as a wedding gift to a couple for them to finish an entire room in their house. Other products from Ed's mill include siding, wood for sheds, and wood for his daughter's garage. Ed's wood has even been shipped to Alaska for use by an artist.

A prerequisite of participating in the firewood club is joining a pre-workday meal of pancakes and sausage.

While Ed very much appreciates being able to supply a necessary and environmentally friendly product, it's not about the money. Instead, he uses wood to connect with people. One of Ed's hard-and-fast rules is that he doesn't deliver lumber. In fact, he doesn't even own a pickup truck. He's happy to supply his home-grown wood for a friend's or family member's building project, but to do so, the recipient must come over and spend the day with him, helping mill the logs and stack the boards. There is also one more important requirement to getting the boards: You have to have lunch with him. The day is an opportunity to connect. "When I'm working on wood and getting it ready for somebody, I feel that I'm actually with that person during that time."

Beyond providing wood products for building materials, Ed has hosted a "firewood club" on his property for over 25 years. Each year, a handful of friends and neighbors gather at Ed's property to help one another fell, skid, buck, split, and deliver enough firewood to meet everyone's needs for the winter. The cast of characters changes from year to year as people drop out while others are pulled in by friends. Each person has a role to play. Some are highly capable at felling trees, while others are better at splitting logs into firewood. Everyone contributes and everyone benefits. Everyone also eats pancakes. A prerequisite of participating in the firewood club is joining a pre-workday meal of pancakes and sausage. The firewood ensures warmth through a cold winter, while the pancakes help ensure connection in a world that seems increasingly disconnected.

Ed always looks for opportunities to meet other landowners, especially those interested in wood products and wood crafts. It is highly educational to hear the solutions of others as they steward their lands and turn logs into lumber, he says. Key landowner groups in New York State include the New York Forest Owners Association and Cornell Cooperative Extension's Master Forest Owner program. "Everyone has different forests, goals, equipment, time, and resources," Ed reflects. "Each person finds ways to achieve their goals within their own unique opportunities and challenges. There is much to be gained from sharing our experiences with one another. Sharing a meal provides the time and connection to facilitate that sharing."

"Nobody got into forestry because they love the look of a cut stump. We love the forest first."
—Jim Finley, State Extension Forester, Penn State University

CHAPTER 11

The Successful Timber Harvest

TIMBER HARVESTING CAN HAVE significant positive and negative ecological, personal, and financial impacts. It should not be seen as an end unto itself but rather as a tool to shape your forest's composition and structure to maintain or diversify benefits. With this potential for positive impacts comes some risk that the harvest, if not done well, will fail to meet your goals for the forest, not pay you fairly, and result in negative consequences for the forest itself, such as rutting of skid roads, loss of financial value, and damaged residual trees.

The fact that this chapter is here should not be seen as an assumption that you should or will do a timber harvest. It is simply a recognition that many landowners choose to implement one, but most lack an understanding of what it entails. If you're going to do it, learn to do it right by asking questions and availing yourself of the many people (professionals and neighbors) and resources available to help you. Remember, this is your land. If you're not certain about the forest management strategy being implemented, the people implementing it, the money you're being paid, the contract you are signing, or anything else, take a breath. Hold off and gather more information; you'll be very happy you did.

A thoughtful timber harvest can be an important tool for achieving your goals.

A thoughtful and well-supervised timber harvest will yield vigorous residual trees of diverse species.

WHAT A TIMBER HARVEST SHOULD LOOK LIKE

To those unaccustomed to timber harvesting, it can be an acquired taste. That's a polite way of saying that even the best harvests can look very messy, not unlike a natural disturbance. There is certainly a drastic visual change between a closed-canopy forest and one with either gaps in the canopy or, depending on the desired light level being created, maybe most of the canopy removed. Nevertheless, it's important to know what to expect and what a conscientious timber harvest looks like.

It can be difficult to see beyond the aesthetic impacts and the magnitude of change to get a true sense of the quality of the job, but it's important to do so. The reality is that while the slash may be unsightly and difficult to walk through for a time, it brings with it ecological benefits. Learning to calibrate your eye to the characteristics of a good timber harvest will help you be a more confident landowner. When evaluating a timber harvest, it can be helpful to break it down into two parts: the overstory and the ground conditions.

Overstory

A timber harvest implementing thoughtful standard forest management will primarily focus on the removal of trees of lower vigor and financial value to increase the quality, complexity, and resilience of the remaining forest. The residual trees in the overstory should include some of the best-quality high-vigor trees in the forest. Crown position is an excellent indicator of a tree's vigor. Trees in a dominant or codominant crown position have proven themselves to be well adapted to the site. Individuals with large, full crowns also have a greater capacity to take advantage of increased growing space than trees of smaller, subordinate crown positions.

Large crowns also provide more area for seed production, which has implications for both regeneration success and food for wildlife. The residual trees should also represent species that will help achieve landowner goals. Increasingly, it's also important to consider favoring species predicted to be well suited to future climate conditions or less vulnerable to insects or disease of concern. Those trees in dominant and codominant crown positions are also trees that have a higher likelihood of living long enough to develop old tree characteristics. So, when evaluating a timber harvest, look up. This is best done after the forester has marked the trees to be cut with paint but before the harvest actually happens. Do most of the trees that will be left on the site have full crowns in a dominant or codominant crown position, or are they suppressed crowns?

Incorporating the concepts of ecological forestry into standard forest management will mean that it is good and desirable to leave not only the straight trees of high economic value but also a diversity of species and some individuals that have old tree characteristics, such as pockets of rot. While these trees are not economically valuable, it took decades or even centuries to develop these characteristics. The number you leave will depend on how much of the overstory you are willing to dedicate to ecological forestry. You must balance your goals.

Warning: If all the trees to be removed are the big, straight ones with the fullest crowns (meaning that afterward your forest is going to be left with poor-quality trees), it's time to put the brakes on the timber harvest and make a new plan.

Ground Conditions

Soils are the foundation of a forest ecosystem. They provide the medium in which trees and plants grow as well as water filtration and tremendous biodiversity, and they are generally the biggest pool of stored carbon within a forest. Well planned and well executed harvests seek to protect these life-sustaining benefits by minimizing negative impacts on the soil, including excessive rutting and soil compaction.

However, some impact on the soil is inevitable. Timber harvesting involves bringing big, heavy machinery into the forest. This means that the soil will not be pristine like in a forest under passive forest management, but it's also not disturbed to the extent that agricultural soil is plowed and worked every year. Forests under active forest management lie somewhere in between, though given the decades of time between most harvests, definitely much closer to the pristine end of the spectrum than the agricultural end.

Understanding what impacts are acceptable and what impacts cross a line to become unacceptable is crucial for evaluating the quality of a timber harvest. Knowing these parameters is similar to obeying the speed limit of a road. At some point, there is a speed beyond which it is unsafe for the driver and others on the road. We are fortunate that our forests are generally very resilient, and a certain amount of impact is acceptable, but we do not want to push the boundaries of this resilience.

Seeding and spreading hay on sections of skid road vulnerable to erosion will help stabilize and revegetate the skid road quickly.

The ground conditions of the timber harvest are the result of decisions that the forester and/or logger have made regarding the timing of the harvest, the amount and location of the skid roads used to transport the logs to the roadside landing, and the strategies used to cross any streams and wetlands. Skid roads should be planned in advance to minimize their number and to locate them on stable ground. Timber harvests should only be conducted on dry, frozen, or otherwise stable ground (such as ground stabilized with slash). The timing of the harvest is critical and, although waiting for the right conditions may be frustrating, temporary pauses in the harvest might be necessary if conditions deteriorate. The types of equipment should also match the site. Choosing the right machine for the harvest will help mitigate impact to the site, particularly as we face the effects of climate change, such as more frequent and more intense storms and winters that don't freeze the ground.

When reviewing a timber harvest, consider whether there is an excessive number of skid roads through the woods. Determine whether the skid roads avoid resource areas such as wetlands and stream crossings and are instead located on dry, solid ground. Look for excessive rutting, tracks from the machines that are more than 6 to 12 inches deep for long stretches. Rutting beyond this depth runs the risk of severing tree roots and mixing soil layers until they are different from how they developed and function best. Finally, at the end of the job, determine if measures have been taken to stabilize the soil and get water off the roads, such as through the use of water bars (small berms of soil running at an angle across the road and into the undisturbed soil). Bare soil and water running down roads leads to erosion. Water bars help, as does the seeding of roads with a conservation mix of native grasses and legumes.

Closing the site, or "buttoning it up," can also include blocking access to the property to deter off-road vehicles and seeding the landing (where cut logs are stacked for transport to a mill). Buttoning up the site properly is critical to its long-term stability and, when done well, can provide benefits such as increased access to your woods for recreation and gathering cordwood and for future timber harvests.

UNDERSTANDING THE PLAYERS

The people you choose to work with and the way your timber harvest is implemented will determine if the harvest will meet your goals, including keeping your forest in good post-harvest condition; ensuring you're meeting relevant local, state, and federal environmental regulations; and protecting your interests. There are several very different professionals involved in forest management. Landowners regularly confuse foresters and loggers, but each has a unique role to play. (Loggers and the different types of foresters are discussed in Chapter 5.) Remember that having a successful timber harvest, or implementing any forest management strategy, necessitates knowing each of these professionals, the knowledge and skills they offer, and whose interests they represent in the transaction.

Often the professional best suited to help you plan your timber harvest is a consulting forester, a trained professional who operates or works for a private business and assists landowners in learning about and making plans for their woodlands. They are paid directly by the landowner and represent the landowner's interest in forest-related activities.

Like a consulting forester, an industrial forester is also a trained forester but one who is employed by a sawmill or other wood-related industry. Industrial foresters may offer the same services as a consulting forester, with the understanding that wood from any timber harvest would be sold to the mill they represent, as their primary role at the mill is to procure the logs to meet its orders. As a result, these foresters are also sometimes referred to as "procurement foresters." This does not mean that industrial foresters and landowners can't share a vision for the forest, but it does mean that the industrial forester is not representing the landowner's interest in the transaction. This is a very important distinction to be aware of when selling timber, both in terms of negotiating the money being offered and for the specifics of the contract. It is always advisable that a landowner have someone representing their interest in negotiating a contract and stumpage values.

In addition to foresters, timber harvesters or loggers are another kind of professional involved in timber harvesting and sometimes in other forest management practices, such as wildlife habitat. A logger is a private contractor who specializes in harvesting, processing, and transporting trees from the woodland. Loggers are experts in safely felling trees and moving them to the roadside landing where they can be picked up for delivery to a sawmill or other wood-processing facility. Every logger has their own harvesting system, a combination of equipment used for felling and removing timber and other wood products from the woods. For example, some loggers may use a chainsaw to fell trees and a cable skidder to drag the logs to the landing, while others may use a cut-to-length system in which trees are felled using a harvester with a saw on the end of a boom and carried out of the woods on the back of a forwarder.

HIRING A CONSULTING FORESTER

A successful timber harvest starts with identifying a good consulting forester who understands your goals and values and can represent your interests. When speaking to a potential consulting forester, be sure to communicate your goals and interests clearly and openly. Hiring the right forester will result in a number of pieces falling into place, such as an accurate tally of your tree volume and quality, a strong contract representing your interests, and a reputable, skilled logger with a harvesting system that matches your woods.

As with finding any professional, the best place to start is with your friends. Who have they used? Who would they recommend? You can also contact the local district/service forester for advice. Because they are public employees, service foresters typically cannot make specific recommendations, but they often have a list of local foresters. Your local land trust may also be a source of information about local foresters. Hiring a forester who lives locally will make it easier for them to supervise timber harvests and other work.

There are a number of professional organizations for foresters, including the Society of American Foresters, Forest Stewards Guild, and the Association of Consulting Foresters. A forester's involvement in a professional organization can be a good sign that they are engaged in their profession and staying current on the latest thinking on forest management relative to climate change and other challenges, such as invasive plants and insects.

It may be that your forest has characteristics that lend themselves to a specific harvesting system. Skidders are better on steep, rocky terrain, but forwarders are better at moving logs a long distance. Timber harvesting machines often cost hundreds of thousands of dollars, and logging requires the ability to remove trees in a low-impact, safe manner, as well as excellent business skills.

As a landowner, you may elect to have one or more of these professionals involved in your timber harvest. The choice is yours. You may choose to sell your stumpage directly to a logger or a mill. This may seem efficient because it removes a middle person from the sale. However, you're eliminating the person who is paid to advocate for you and to ensure that the harvest achieves your goals and that you are paid a fair price; that you meet local, state, and federal laws pertaining to timber harvesting; and that the forest is left in good shape post-harvest.

SELLING STUMPAGE

With an understanding of the players involved in a timber harvest, let's discuss the common ways in which stumpage is sold. To be clear, you are not "hiring" a logger; your trees are worth money, so you should be paid for your timber harvest. It is very common, and very reasonable, for a landowner to want to know the "fair" price they should be paid for their stumpage. After all, you may sell timber only a few times in your life, so you might not be familiar with considerations like

WHAT IS STUMPAGE?

Stumpage is the price paid to the landowner for their standing trees and the right to harvest them. It does not include the time and expense of felling the tree and transporting it to the roadside for delivery to a sawmill.

Stumpage can be sold in several ways. The professionals you choose to work with and the way you sell your wood products will have a significant impact on the type of management implemented, the condition in which your woods are left, and the financial value of the harvest. Upon selling stumpage, the landowner grants access to a timber harvester to remove the trees they have bought. The specifics of that access, such as a description of the specific trees that will be removed, the dates within which the logger will have access, the ground conditions under which the access may be granted, the schedule of payment to the landowner, and the condition in which the forest will be left after the harvest should all be detailed in a contract written to represent the landowner's interests. See page 237 for the important elements of a contract.

the value of timber and your options for selling it. Learning more about the factors contributing to stumpage prices will help you obtain a fair price.

The Value of Your Timber

Many factors affect the value of your timber, including the tree species, size, and quality and the volume to be harvested. Other factors, such as terrain, skidding distance, and the presence of streams and wetlands in your woods, affect the difficulty of the logging job and therefore the price that a timber harvester may offer. Landowners can become frustrated when they receive different offers from different loggers. Each logger has a unique combination of harvesting equipment, markets for the trees, and costs for operating their machines. Therefore, even if they are proposing to harvest the exact same trees, different loggers are likely to offer different prices.

Some loggers may have harvesting equipment that fits your forest very well, allowing them to harvest quickly and efficiently. Others may have machinery that will do the job but less efficiently. One logger may have a niche market for your trees to produce a specialty product that allows them to pay you more, while another may be selling the same trees for a low-grade wood product. Some loggers may not be busy and are therefore eager for your business, while others have two years of harvests scheduled. As a result, variation in prices shouldn't always be seen as dishonest but rather as reflecting the unique circumstances of each logger.

However, you can't be naïve—a logger may simply be offering you a price that is lower than the value of your stumpage in the hopes of getting the wood for as little money as possible, a circumstance you would do best to avoid. Ultimately, ensuring a fair price necessitates having an accurate estimate of the volume and quality of the trees to be removed, knowledge of the current prices for stumpage, and experience in determining the market value of the stumpage. This is one of the ways in which a consulting forester, who is representing your interests, can help.

Marking Trees

If you choose to work with a consulting forester, they will likely discuss your goals for the timber harvest. Your forester will determine the best strategy to create the forest conditions to meet your goals. They will then mark the individual trees to be removed, using paint. There are two advantages to having a forester mark the trees to be removed, one ecological and the other financial. Foresters are trained to understand the ways in which forests can be changed to encourage the desired species composition and structure to meet landowner goals. When a forester marks the trees to be harvested, you are having someone with formal training and experience make these decisions. As a forester marks trees, they will tally the species, diameter, height, and quality to come to an estimate of the tree's volume as well as determine its product class.

The way the tree is marked often communicates the type of product for which it was tallied. For example, a horizontal line at breast height (4.5 feet off the ground) often means that the tree was tallied as a sawtimber tree, while a slash at breast height means pulp, and a dot means firewood. You should also see a painted dot on the stump of the tree. After the tree has been harvested, the stump spot verifies which trees were indeed meant to be removed. Once all the trees to be removed are tallied, a forester can estimate the total value of the sale based on their knowledge and experience of the current market condition. This estimate can be used when selling timber.

Marking each individual tree is the best way to ensure the ecological and financial benefits of a timber harvest, but it is time-consuming. There are circumstances in which a forester does not mark all of the trees to be harvested. For example, a forester may *reverse mark*, putting paint on all the trees to be retained. There also may be situations in which a forester works with a logger to develop a harvest prescription, describing the species, size, and qualities to be removed. A landowner's leverage in the negotiation is at the front end, before the trees are harvested. Whichever manner of tree selection and volume estimation is decided on, be sure you are comfortable with it. When searching for a forester, talk to them about their approach when setting up a timber

Having a forester mark the trees to be removed ensures that you are benefiting from the forester's expertise to create the desired conditions, and provides the logger with a clear picture of which trees should be removed.

harvest and be sure to communicate your preferences.

Determining a Price

Once the trees to be harvested are selected and the volume is estimated, there are three general ways to sell your stumpage: with a negotiated price, a competitive bid, or a mill tally.

Negotiated price. A price is negotiated for the value of your timber. The negotiation can be done by you or by someone with a greater knowledge of timber prices representing your interests, such as a consulting forester. The money is often paid as a lump sum (a single payment in advance for all of the timber offered for sale) or in a couple of installments (such as half up front and half when the harvest is halfway done). Negotiated prices are most frequently used when timber harvesters or industrial foresters buy timber directly from landowners. Consulting foresters will also use negotiated sales for timber harvests of marginal value or when specific equipment or expertise is needed. The advantage of using a consulting forester to negotiate the price paid for your stumpage on your behalf is that they have an accurate accounting of the volume to be removed and its quality, allowing them to develop a fair price based on the current market.

When a consulting forester negotiates a price with a logger or sawmill, two parties of more or less equal knowledge are determining the fair price. That's the free market system. A landowner negotiating their own prices with someone of far greater knowledge is not the free market; it's predator-prey.

Competitive bid. A competitive bid is used by consulting foresters to determine the value of your timber in the open market and can give you the most money, especially for high-quality timber. The trees to be cut are marked in the woods with paint, and their volume and value are estimated. A "timber showing" is organized, and responsible timber harvesters with appropriate harvesting systems are invited to the showing by the consulting forester. Bidders visit your property, assess the timber, and evaluate the difficulty of the logging job. They are then asked to submit sealed bids by a certain date. The consulting forester and landowner open the bids together, consider the prices being offered, and decide which bidder to award the job to. Although the highest bid is often awarded the job, the landowner may reserve the right to refuse any or all bids in case there is compelling reason to award the job to someone who wasn't the top bidder or the money being offered isn't to the satisfaction of the landowner and forester. The price for timber can vary considerably. A study done in Massachusetts showed that, on average, the highest bid is more than twice as much as the lowest, demonstrating the wide spread of prices people may offer for your stumpage.

Mill tally. Both the negotiated price and the competitive bid approach are strategies for selling stumpage—that is, standing trees before they have been harvested. Rather than selling stumpage, it is possible to sell the logs themselves through a method called mill tally. In the mill tally method, a landowner and logger agree in advance on a price for each species. The trees are then harvested and taken to a mill where they are measured and tallied by species in the mill yard. The mill produces a tally sheet of the logs, their species, and the volume. The pre-negotiated price for each species is then applied to the tally to determine the money paid to the landowner.

For example, if the negotiated price for white pine was $100 per thousand board feet (abbreviated MBF) of volume and the white pine logs brought to the mill from your forest totaled twenty thousand board feet (20 MBF), then you would be paid $2,000 ($100 × 20 MBF = $2,000). The advantage of the mill tally method is that tallying the logs in the yard allows for a more accurate measurement than a field-based estimate when the tree is on the stump. However, the success of the method demands that all of the logs are accounted for and all of the mill tally slips are paid out. This can be difficult or impossible for a landowner to know. As stated above, when using the negotiated price and competitive bid methods, the price is determined up front. By contrast, the mill tally can be seen as a loss of leverage.

Be sure you understand how the value of your stumpage is being determined and that you are comfortable with it. This is your forest! Don't get pressured into a deal that you are uncertain about. If you have any doubts, hold off and collect more information.

Understanding the nuts and bolts of a timber harvest, including the professionals involved, the roles they play, and your options for selling stumpage, are all part of implementing a successful timber harvest. Once these important decisions have been made, formalizing the players and plans for the timber harvest in the form of a contract is a necessary next step to protecting your interests and achieving your goals.

Common Elements of a Contract

Like many things in life, it is crucial to have agreements in writing. It is therefore always advisable to have a written contract when selling stumpage. Be sure that any timber harvest contract you sign represents your goals and interests. A consulting forester representing your interests will likely have a template for a contract that can be tailored to meet your needs. Others will likely have a contract that represents theirs. Below are the common elements of a forest management contract.

Statement of Buyer and Seller

Identify who is selling the timber, who is buying it, and, if applicable, any agent acting on behalf of the seller. This would be a place to clarify who might be administering the sale. For example, if you choose to work with a consulting forester, your consulting forester can be named as your agent for the timber sale with authority to make decisions about the harvest. This is especially important if the seller is not going to be present or lives out of state.

Dates of Operation

This section of the contract should identify the total duration of the contractual agreement as well as the consequences if the work is not completed (for example, the contract expires but not all of the timber has been cut or removed from the property). This is also a good place in the contract to address the possibility of needing to temporarily shut down the job due to weather or ground conditions and the person responsible for making that decision, preferably someone representing your interests, such as your consulting forester.

Always record the date that the contract is signed by both parties. Contracts often run for as long as two years. This actually allows for the possibility of needing to shut down due to unfavorable conditions. It's common for landowners to want to create a short window for the timber harvest to get it done, but a short window can create problems. If the timber harvester feels rushed to finish on time, regardless of the weather or ground conditions, it can lead to very undesirable impacts to your forest. A longer contract period allows the timber harvester to operate during stable site conditions, protecting your forest and giving you better results.

Units of Measure

There are many ways to quantify the volume of wood to be harvested. These units can vary by wood product and by region. A *log rule* is the mathematical formula used to estimate the value of wood in a log. There are well over one hundred log rules in use in the United States today. Which units will be used to estimate the volume of wood to be removed from your forest? Cubic feet? Board feet? Tons? Cords? Cubic meters? And if board feet, which log rule? International ¼ inch? Scribner? Doyle? Your consulting forester can help you understand the unit of measure used in the contract.

Price and Payment Schedule

This is the place in the contract where the price is specified, along with the timing of payments. How will the payments be made? They could be paid on a lump-sum basis—for example, $10,000 for all timber marked with blue paint, or $47,415 for all timber within the area designated on the map as "lot A." Or they could be paid on a mill tally basis, as estimated by the mill that receives the logs. For example, $300 per MBF of red oak, $40/MBF of red maple, $120/MBF of white pine.

Performance Bond

Sellers of standing timber commonly require a security deposit from the buyer. It could be in the range of 5 to 20 percent of the full value of the sale. By holding the security deposit, the landowner selling the timber has some insurance that all provisions of the contract will be met, particularly those that leave your harvest area in good shape, such as meeting all required BMPs, smoothing the roads, seeding the landing, controlling erosion, and lopping slash. Of course, this part of the contract should also specify the conditions for the full return of the security deposit. If your timber harvester doesn't meet all provisions, you may end up needing the money to pay someone else to clean up the work.

Insurance

This is very important. If there are problems on a timber sale (a logger is injured, a curious neighbor child wanders onto the harvest site and is injured, falling timber causes damage or worse), it is possible that the landowner could somehow be seen as potentially liable. For this reason, it is an excellent idea to require that the buyer of the timber hold insurance for property damage, bodily injury, and workers' compensation.

Simply stating in the contract that the owner is not liable may not be sufficient in the eyes of a court. It is advisable, however, to include such a "save harmless" clause in the contract.

Harvest Tree Designation

Specify the way in which trees to be harvested are designated in the woods. Will the timber be marked? What color paint will be used? Will the area be designated with ribbons or paint? Will there simply be a description of the trees to be harvested?

Harvest Map

It is advisable to include reference in the contract to a map designating the area on the property where the harvest will take place. The map should indicate locations of landings, stream crossings, boundary lines, and critical areas to avoid (such as wetlands and sensitive habitat).

Landing

A landing is the location on a logging operation to which logs are transported from the woods, stacked in piles according to their product class, and then loaded onto log trucks for transfer to a mill. Landings are usually located near public roads, and skid trails lead out from them into the woods. Logging equipment such as skidders or forwarders travel over the skid trails to bring the timber from the stump to the collection point for transport. As harvesting systems have grown to include more equipment, the size of landings has grown to accommodate the equipment. You might be surprised by the amount of area necessary for landings. Be sure to include the size and location of the landing in the contract.

It is advisable to find a forester who will represent your interests in the timber sale and develop a strong contract to protect those interests.

Post-Harvest Condition of Roads and Landings

A good network of woods roads greatly facilitates timber harvesting by providing stable access to the forest. Beyond the operational considerations, woods roads can provide recreational opportunities such as hiking and snowmobiling. They also give landowners opportunities to harvest firewood and can serve as fire breaks. To protect all the benefits woods roads and trails provide, it is important to state what the woods roads conditions are prior to the start of the harvest to avoid misunderstandings about how they will be left after the sale. Whose responsibility is it to repair them? What about trash or debris left on the landing? In addition, the contract should state any expectations concerning upgrades or improvements to the roads.

Regulatory Responsibilities

Water resources are an important and common part of the forests of our region. Timber harvesting involves bringing big machines into your forest. Taking every reasonable precaution to ensure the timber harvest doesn't negatively impact these resource areas is a critical part of maintaining forest health and a vital part of a successful timber harvest. Beyond the ecological impacts, as the landowner you are responsible for ensuring all local, state, and federal regulations are met on your land. Be sure you have a clear sense of the relevant laws and regulations that the timber harvest must follow. Most laws and regulations focus on preventing nonpoint source pollution from timber harvests from entering regulated resource areas such as wetlands, streams, lakes, ponds, and vernal pools. Some states may have other regulations, such as acts regarding forest-cutting practices that specify regeneration requirements and/or protect important habitats. Be sure that all regulations will be properly met.

Slash Height and Location

Slash is the debris that is left after harvesting. It is the branches, limbs, and other pieces of trees left in the woods. Many owners believe it is very unsightly and are concerned about the apparent "mess" that slash makes during and after the harvest. It is possible to specify in the contract that slash on the ground not be left higher than a certain level, or not left at all within an agreed-upon distance from trails, roads, boundaries, homes, or other features. Cutting slash by hand is dangerous, expensive work and will reduce the stumpage value.

Keep in mind that slash provides many ecological benefits. It provides habitat, adds nutrients to replenish the soil, and protects new regeneration from browse. If you can overlook the short-term unsightliness of slash, you may end up with a contract that is more economically and ecologically desirable.

Penalty for Cutting Unmarked Timber

State and local regulations on timber trespass—harvesting trees without permission—can include civil and criminal penalties. The contract should specify a penalty for cutting timber that was not designated for harvest.

A strong contract can help ensure that your timber harvest achieves your goals and leaves your forest in good shape.

Commonly, timber that is marked for harvest is painted both at eye level and at ground level. Thus, paint should be evident on the stump to indicate that a designated tree has been removed. Stumps without paint imply that the timber was not designated and agreed upon in advance.

Improvements to the Woodlot

A timber sale is an excellent opportunity to have additional work done on the property. For example, an extension of an existing road could be built with the equipment used by the logger. Gates could be constructed, bridges built, access otherwise improved, or other timber improvement practices implemented (such as precommercial thinning). Timber harvesters are in an excellent position to make improvements like this since they have heavy equipment and experience. These kinds of improvements must be included in the contract, however, so that all parties understand the expectations and the timber is priced accordingly. As in contending with slash, these additional improvements cost the buyer time and money. Sellers should expect the price of timber to be lower if additional improvements are added.

Establishing a Working Relationship

A strong contract that represents your interests is a critical part of a successful timber harvest. However, a strong contract must be paired with a good working relationship with your forester and the logger implementing the harvest. You each have an important and complementary role, and cultivating a respectful relationship with open communication and collaboration is key to a successful harvest.

In some ways, the contract is a last resort to be used when all else fails. Forests and, increasingly, our weather are dynamic and unpredictable, which can lead to unexpected issues. It could be the ground looked stable but wasn't. Or a large rain event might happen. In fact, it's likely that an unexpected issue will arise. Identifying the issue early and working as a team to address it to everyone's best interest is key.

GETTING STARTED

Timber harvesting can be a powerful tool. Your goals for your family and your land should dictate the type of harvest you implement in your woods. All harvests do not provide the same benefits. Know your options and their likely outcomes. Working with a consulting forester is an excellent way to evaluate your options and plan an approach that will fit both your immediate and long-term goals. Finding a good forester who understands your goals and can represent your interests is well worth your time. A carefully planned timber harvest will address your financial, wildlife, aesthetic, and recreational goals while meeting your regulatory responsibilities.

Above all, don't rush. Move forward only after you have enough information to make a decision that is right for you and your family. A decision to postpone a harvest or not to cut timber at all has the benefit of maintaining your management options into the future. A high-grade harvest that removes only the valuable trees or a poorly implemented timber harvest will limit your future management options and reduce your enjoyment of your forest.

Practicing Ecological Forestry for Timber Harvesting

CONTINUITY: When planning any type of timber harvest, always focus on the elements of the forest that will be retained, including deadwood, legacy trees left to develop old characteristics, diverse species, and trees around environmentally sensitive areas and unique habitats. The amount of retention should be decided in conversation with your consulting forester and in consideration of your overall goals. Contracts should specify the protection of these elements. The layout of the timber harvest should avoid damage to them.

COMPLEXITY AND DIVERSITY: Harvesting wood products shouldn't be the endpoint of a timber harvest. Instead, a harvest should strive to create structural complexity and species diversity. Examples include creating the light conditions necessary to establish regeneration of diverse species adapted to future conditions and increasing the spatial variation of the trees.

TIMING: Rather than deciding the timing of timber harvests based on financial maturity (the point at which a tree's rate of value increase falls below a desired level), allow the forest, or at least parts of it, the time necessary to develop the characteristics missing from our forests, including structural complexity, deadwood, and old-tree characteristics, such as complex bark and pockets of rot.

CONTEXT: We are all wood consumers. That wood needs to come from somewhere. When we don't harvest locally, we're exporting our responsibility to others, often at carbon, ecological, and ethical costs. The solution is to produce wood locally through ecological forestry. Understand the important role you play in your community, region, and country in helping to meet our wood product needs, and be proud of it.

"In every deliberation we must consider the impact of our decisions on the next seven generations."
—Seventh Generation Principle, Haudenosaunee Confederacy

CHAPTER 12

The Legacy of Your Land

CONTINUITY IS THE FIRST PRINCIPLE OF ECOLOGICAL FORESTRY, and for much of this book, we have focused on ecological continuity. However, there is an equally important social context. Ensuring a continuity of forest cover from your ownership to future ownerships, and providing for the time forests need to develop their wonderful complexity, requires conservation-based estate planning. The decisions you make about the future of your land will have profound impacts on your forest and the benefits it provides (or doesn't!). In fact, formally planning for the future of your land may be the most important step you can take as a landowner—not just for your benefit but for the benefit of your family, your community, and the land itself. Who will own your land, and how will it be used? What will your legacy be?

Determining the future ownership and use of your land is essential to reaching your personal, financial, and ecological goals.

WHY HAVE AN ESTATE PLAN?

An estate is the total of all your assets, which includes your land, house, and bank accounts, as well as any stocks and bonds. An estate plan ensures that your assets will be distributed in a way that will meet the financial and personal needs of you and your heirs. Although the phrase *estate plan* may bring to mind a single all-encompassing document, it is best thought of as a combination of documents (such as a will) and tools (such as a conservation easement) that will achieve your goals when implemented together. Estate planning is not just for the wealthy or for those who own "estates"; if you own land, then estate planning is a necessary and valuable step for protecting your legacy.

Every landowner and every family is different. Some have children who are prepared and excited to become the new owners of the land, while others are searching for creative solutions to relieve the financial pressure and responsibility of owning land. Landowners who do not have children may be seeking a way to maintain the land as forestland and want to find a suitable owner.

Successful estate planning will often avoid certain taxes, increase the assets given to your heirs, address your family's goals for owning and using the land, ensure financial security for you and your family, and maintain good family relationships. Failure to plan for your land's future means giving up control of its future ownership. It also may result in negative financial consequences, may allow the land to be treated in a way counter to your goals, and may lead to tension or animosity among your family members, which can last long beyond your passing.

Your land is likely one of your most valuable assets, especially if you have owned it for a long time and it has greatly increased in value. However, land is not like other assets. Because land can be connected to memories, experiences, and feelings not typically associated with other assets, such as stocks and bonds, your land may also have significant personal value.

Dividing your assets among family members brings with it the challenge of providing for their financial as well as personal needs. If you have family, some of them may be interested in receiving financial reward from the land or obtaining a piece of land on which to build a home. Others may be most interested in keeping it in the family and in its current or natural state. And, of course, it is possible that family members may want or need a little of both.

The good news is that land is a flexible asset that lends itself to creative solutions for gaining financial and personal value. Whether your intention is to keep the land in your family or not, it is possible to develop a solution to meet your needs and goals as well as those of your family. Unfortunately, there are countless examples of landowners who put off a decision about their land until it's too late. The earlier you start your planning process, the more options you will have for your land.

CONSERVATION-BASED ESTATE PLANNING

This chapter focuses on conservation-based estate planning, with the goal of keeping some or all of your land in its natural, undeveloped state. Most people understand their options to subdivide and develop their land. However, many are not aware of their land conservation options, the variety of helpful legal tools used for transferring land, and how using these tools separately or in combination may produce a result that better meets one's personal and financial goals.

ARTICULATING YOUR GOALS

Begin by thinking about what you want most for your land and family after you are no longer around: To see the land remain undeveloped forever? To maintain family harmony? To ensure financial security? To preserve flexibility for your heirs? You may want to rank your goals in order of their importance and list any challenges standing in the way of those goals.

If you own your land with others, your next step will be to talk about your goals with your spouse or co-owners and try to incorporate shared goals and values as you lay the foundation for your plan for the future of the land.

Involving Family in the Decision

If you have children, it's up to you to decide to what extent you want to involve them in determining the future of your land. Every family's situation is different. Involving your family can help avoid conflict and get their buy-in. In planning for the future of the land, families may be forced to face difficult subjects, such as aging, death, and how family assets are to be distributed.

Discovering your family members' needs and wishes can take place either at a family meeting or through individual conversations when there is a good opportunity to talk. Be sure to choose the strategy that best fits your family's style. Treat these discussions as an occasion to invite questions and discuss options. In doing so, you're also creating a protocol for honest, productive, and respectful family communication.

A family meeting can be a useful forum for giving family members a chance to share their ideas and concerns and hear the needs and wants of others. This can be accomplished by simply asking each person to talk about how they feel about the land. Is it a priceless family heirloom to be protected at all costs? Is it a financial asset and nothing more? Or is it something in between? By listening to one another, family members may learn that they share certain feelings—or, just as important, that there are differences. This information can guide your next steps and inform your work with estate planning professionals. It may be helpful to have an estate planning professional attend a family meeting to provide technical information and answer questions directly.

Sometimes a family's history or dynamics prevent them from having healthy conversations about what to do with the land. However, avoiding these important conversations and letting your family figure it out after you are gone will likely lead to even more tension. A neutral person or professional facilitator can often help guide these difficult conversations; alternatively, simply having individual conversations with family members can be a good strategy.

Though your goal may be to get your family to agree on a plan for your estate, there may be situations in which families are not able to work together or agree. In this case, you need to be prepared to take the input you have received, work with the necessary professionals, and do what you believe is right for yourself, your family, and your land. Do not become paralyzed by disagreements. If you avoid planning because people don't agree now, you can be sure the conflict will be greatly exacerbated after you are gone.

After you have determined your goals for the land, maintaining momentum is essential. Important questions may have come up in your discussions. Use these questions to help guide your next steps. Contact a professional to help you move your goals forward.

Equal vs. Equitable Treatment

From the time their children are very young, most parents strive to treat each child equally. When it comes time to plan the future of your land, it's important to realize that it's possible to treat your children equitably, or fairly, without their inheritances being precisely equal. You may not have enough assets to give your heirs shares of equal value, and different heirs may be carrying more of the management responsibility or costs for the land. If you have an honest conversation and set up a fair process to determine your children's needs and values, you can still treat them equitably without dividing all your assets into equal shares.

Perhaps other assets, like a life insurance policy or proceeds from the sale of a house, can go to one child while the land goes to another. Maybe selling a house lot will provide enough income for one heir so that another can receive the land. Keep in mind that giving your land to each heir in equal parts can set the stage for the undesirable outcome of further dividing or selling the land to generate enough income for one uninterested heir.

DETERMINING THE FUTURE OWNERSHIP OF YOUR LAND

Ask yourself if you want to determine the future ownership of your land. If you don't, you can sell it and use the proceeds to achieve other goals. If you do want to determine who will own your land, there are conservation-based estate planning tools that will help you formalize your wishes.

Using a Will to Determine Who Will Own Your Land

A will, also called a last will and testament, is the most commonly used estate planning tool. A will is a legally binding document that states how you want your assets distributed once you have passed away. Wills are simple, inexpensive ways to address many estate planning needs, but they can't do it all. For example, a will won't help you minimize taxes or avoid probate. Probate is the winding down of your affairs under the supervision of the courts. Since probate makes all your assets public information, some people want to avoid it. However, for families that are dealing with conflict or miscommunication, it can be beneficial to have the courts involved. To be effective at passing your land via a will, you need to ensure that your land ownership permits this—and, if so, that your will is written in a way that will achieve your desired goals. Following are some considerations for the development of a will that effectively distributes your land assets to meet your goals. These suggestions are not meant to replace the advice of an estate planning professional but rather to help guide your conversations with one.

Description of current owners. If you own only a partial interest in the land, or if you want to make a provision for how the land will pass from the survivor of joint owners to the next owner or owners, it is helpful to state your ownership interest in your will.

Beneficiaries. Your will should include the names of the individual(s) (spouse, children, siblings) or organization (land trust, state conservation agency, local town) to which you would like to give your land. If you do not have heirs and are interested in giving your land to a land trust, state conservation agency, or local town, it is highly recommended that you meet with the organization and discuss your intentions. Though you might think that any organization would jump at the chance to own your land, be sure to have in-depth conversations with your chosen recipient so that nothing comes as a surprise.

Avoid conditions. It is not advisable to set up conditions for the ownership of the land—for example, "My daughter can have the land as long as she doesn't build houses on it." Conditions can be difficult both to interpret and to enforce and thus may not ensure the outcome you want. In addition, wills are intended only to transfer your assets according to your wishes and are therefore meant to have a limited lifespan. If your goal is to keep some or all of your land undeveloped, it is recommended that you consider placing a conservation restriction on the land or giving the land to a conservation organization, such as a land trust or a state conservation agency.

Using the Type of Land Ownership to Determine Future Owners

A great place to start your conservation-based estate planning is by reading your deed, which lists your current form of ownership, and understanding the implications of this form of ownership. You can find a deed in your local registry of deeds or land recorder office. These records are increasingly being put online. An estate-planning attorney, a land trust, or a forester can also help you find your deed.

Without estate planning, the form of ownership listed on your deed will determine who will get your land when you pass away—and it might not be what you had in mind for your land. If your current form of ownership doesn't meet your needs for passing on your land, consult with an estate-planning attorney to learn about other options.

If you do want your family to continue owning the land, you will need to choose the form of legal ownership that determines who controls the land, how it is transferred, how it is taxed, and how liability will be shared, among other things. Determining which form of ownership is best for you depends on several factors, including the number of people who will be sharing ownership, liability concerns, how income from the land is taxed while you are alive, and how the land will be taxed when it is transferred. There is a range of land ownership options. Bringing your goals, or those of your family, to an estate-planning attorney with land conservation experience is a great way to sort out which type of ownership will be the best fit.

Direct Ownership

Direct ownerships are those in which a person or persons own the land directly. The names of the people who own the land are on the deed. Direct ownerships are the most common form of land ownership. They are easy to set up and maintain, and all forms of direct ownership can be combined with land conservation tools. However, they do not provide protection for liability or a mechanism for the gradual transfer of land. Following are some examples of direct ownership, along with short descriptions of how land owned in these forms is passed on.

One individual owner. Ownership of the land is by a single person, whose name is listed on the deed. Upon the death of the owner, the land is transferred according to the terms of their will.

Joint tenant with right of survivorship. Ownership of the land is by two or more persons, and the deed states that they own jointly or jointly with right of survivorship. Upon the death of an owner, the land automatically passes to the surviving owner or owners. This automatic transfer overrides the will.

Tenants by the entirety. Ownership of the land is by spouses (and spouses only), and the deed states that they own as tenants by the entirety. Upon the death of a spouse, the land automatically passes to the surviving spouse. This automatic transfer overrides the will.

Tenants in common. Ownership of the land is by two or more persons, and each may own specific shares. If a deed does not state that the persons listed own jointly or as tenants by the entirety, then they own as tenants in common. If one person dies, their share passes according to the terms of their will, creating exponentially larger and more complicated ownerships as each owner's share gets passed on to multiple heirs. As a result, making decisions about the land can become extremely difficult.

LAND OWNERSHIP AGREEMENTS FOR MULTIPLE OWNERS

Land ownership brings with it many benefits (recreational use, income from forest management) and responsibilities (taxes, maintenance). These benefits and responsibilities can be complicated when there are multiple owners in a direct form of ownership. To reduce the likelihood of disputes, landowners who co-own land may want to consider a land ownership agreement. A land ownership agreement is a contract between two or more persons who own a piece of property together. The agreement typically describes the following expectations and terms for each party.

- Right to use the land
- Obligation to pay taxes
- Responsibilities for maintenance of the property
- Entitlement to profits
- Transfer rights in the property

A land ownership agreement can outline the rights and responsibilities of owners in an effort to avoid conflict and increase the enjoyment of ownership. It can be a legal document drafted by an attorney, or something simple that documents how a property is to be used as well as expectations for sharing costs and benefits to avoid disagreements over time.

Indirect Ownership

Indirect ownerships are those in which land is owned by a legal entity rather than by a person. Indirect ownerships can be relatively complex to set up, but they do bring with them opportunities to protect individuals from liability, as well as to gradually change or transfer the ownership, which can offer important tax benefits. Following is a list of some of the most common forms of indirect land ownership that families can use to achieve their goals.

Limited partnership. A limited partnership is an organized entity for which income and loss go on the owners' personal tax returns. Taxes and accounting for limited partnerships are relatively simple, but they do require annual reporting to the IRS. Ownership is transferred by interest instead of through a deed, but when one partner dies, the partnership usually terminates unless the remaining partner or partners agree to maintain it.

Trust. There are many kinds of trusts and many reasons for setting them up. In general, trusts operate in a top-down fashion in which the creator, usually the landowner, names a trustee who manages the assets and must act as directed and in the best interests of the trust's beneficiaries. The beneficiaries are those named in the trust who will be receiving assets or who will otherwise benefit from the trust. The way in which the land will be transferred, the timing of the transfer, and how decisions will be made about the land can be written into the trust.

Other common reasons for creating a trust include avoiding the cost and delays of probate, assigning a trustee to run your affairs if you are unable to do so, reducing or eliminating estate taxes, and keeping your estate private. Your state may limit the number of years that a trust can stay viable, so while trusts can offer a good option for families, they are not a permanent solution. The trustee has a responsibility to protect the financial value of the assets in the trust, which may make it difficult to permanently conserve land in a trust, since permanent conservation naturally reduces the value of the land. If permanent land conservation is a goal now or in the future, landowners should include that goal as part of the organizing principle of the trust.

Limited liability company (LLC). When land ownership is transferred to an LLC, the rules that run it are put into an operating agreement and agreed to by all initial members. The LLC is governed by a manager for the benefit of all members. Shares of ownership can be transferred between members by gift or by sale, as determined in the operating agreement. The LLC can be used to gradually move ownership from one person to another or between generations. This gradual transfer of ownership can help minimize taxes and provides a mechanism for accommodating the changing needs and life circumstances of members.

As the name suggests, limited liability companies protect all owners from liability. Unlike trusts, LLCs can be maintained in perpetuity and amended by members over time, providing a long-term strategy for land ownership and decision-making. In addition, land owned as an LLC can be put into permanent conservation. It should be noted, however, that there is an annual cost for maintaining an LLC.

DETERMINING THE FUTURE USE OF YOUR LAND

Would you like to control the future use of your land to ensure that some or all of it remains in its natural state? You can choose to determine the future use of your land either temporarily or permanently. A forest management plan created by a consulting forester can help guide the stewardship of the land in the future. Current use programs are an example of a temporary option for determining future land use. Permanent land conservation, on the other hand, can ensure that your land will never be developed or subdivided.

Temporary Land Use Options

Paying the property taxes on the land may be an issue for your family. As described in Chapter 10, current use tax programs give landowners an opportunity to significantly reduce their property taxes in exchange for keeping land undeveloped and producing

TRUST VS. LAND TRUST

It is common for people to confuse a trust and a land trust. A trust is a legal entity you create to achieve your estate-planning goals, whereas a land trust is a nonprofit organization that helps landowners reach their conservation goals. A trust does not protect your land unless provisions for this are specifically spelled out in it.

public benefit. Although these programs do not typically provide permanent protection for your land, they make owning forest and farmland more affordable and can be used in combination with other land conservation tools. In addition, some states place a lien on the property when the landowner enters the program. The lien can be removed with the payment of back taxes, but it does provide a temporary form of protection to the property that can transition the land from one owner to the next while keeping it in forest use. Contacting your local service forester, extension forester, or a consulting forester is an excellent way to learn more about your specific current use tax programs.

Permanent Land Use Options

There are a few different ways to conserve your land permanently. Consider which one might be right for you and your family.

Conservation Easements

A conservation easement (CE; known as a conservation restriction in Massachusetts) is a legal agreement made with a qualified conservation organization, such as a land trust or a state natural resources agency, that extinguishes some or all of the development rights of the land forever but allows other rights—such as farming, forestry, and recreation—to continue, all while maintaining your ownership of the land. A conservation easement can be placed on all or designated parts of your land, allowing you to reserve house lots to provide financial value or housing options for your family. Some CEs allow public access, whereas others do not; this usually depends on your preference, which organization you work with, and whether you are receiving public funds for your CE.

A CE can be sold to the qualified conservation organization if the land has exceptional ecological or historical value. Alternatively, it can be donated, providing the landowner with a tax deduction for a charitable gift. Since the land can no longer be developed, a CE lowers its value, which can help lower your taxable estate, possibly even dropping it below the estate tax threshold. Donating a conservation easement can also reduce income tax burden. In these cases, landowners are required by the IRS to have the land appraised by a qualified independent appraiser to determine the value of the deduction. Both sold and donated CEs often come with costs for surveys, appraisals, and stewardship donations to ensure that the terms of the easement are monitored and enforced in perpetuity.

Your heirs can donate a CE within nine months of your death, which can provide tax benefits to your estate, meeting the goals of keeping some or all of your land in its natural, undeveloped state and providing financial benefits to your heirs. As previously discussed, before you include the donation of a CE in your will, be sure to meet with the conservation organization you have in mind to confirm that it is willing to accept the CE.

ARTICULATING GOALS IN A CONSERVATION EASEMENT

There are several sections within the legal document of a conservation easement, including the purposes, prohibited uses, and allowable uses. The **purposes** section describes the public values that your land provides that are being protected through a conservation easement. The purposes section also serves as a guide for determining uses that aren't specifically mentioned within the CE. If you are donating some or all of the value of the conservation easement and receiving tax benefits as a charitable gift, these public benefits help justify the charitable nature of your gift.

The land uses that would diminish or eliminate the conservation value of your land in the area of your property covered by the CE are described in the **prohibited uses** section. The most common prohibited uses are forest conversion for residential development and mining. For landowners interested in designating some or all of their land as an area of passive forest management, a description of the location and a definition of the passive approach would be detailed in this section.

Practices that can continue within the area covered under the CE are listed in the **allowable uses** section. If you are interested in allowing active forest management on some or all of your forest, you would clearly describe forestry as an allowable use. Importantly, doing this doesn't require you or future landowners to implement active forest management. Instead, it simply reserves the option. There are opportunities to include the standards of future forest management within the CE, such as implementing forestry best management practices, developing a forest management plan, or using a professional forester.

The conservation agency or organization you choose to work with will help you think through the specific language of your CE. You should also have an attorney who represents your interests, one with experience in land conservation, review the CE on your behalf.

Donating or Selling the Land

Land can be permanently protected by donating it or selling it to a qualified conservation organization, such as a land trust, a state conservation agency, or a town. A donation of land may provide significant tax advantages as a charitable gift.

Bargain sale. Landowners can sell their land or conservation restriction for a price lower than its fair market value. The difference between the appraised market value and the sale price to a qualified conservation organization, such as a land trust or a state conservation agency, is considered a tax-deductible charitable contribution, providing some income and potentially some tax benefits.

Bequest. A donation of land or a conservation easement through your will is another way to ensure your land's permanent protection and potentially reduce your estate tax burden. You can change your will at any time, and a bequest does not become effective until your death. This is a good approach if you need to keep the financial value of your property in reserve in case of unexpected medical bills or other needs, but you want to be sure the land will be conserved if you do not need to sell it during your lifetime.

Life Estate

Landowners sometimes negotiate a gift or sale of the property to a conservation organization while reserving the right to occupy and use the land for life. In these cases, control of the property automatically transfers to the conservation organization upon the death of the landowner. The gift of a property with a reserved life estate can qualify the donor for a charitable deduction based on the value of the property donated and the value of the reserved life estate, which is based on the donor's age. Landowners are responsible for upkeep and all management costs during their lifetime.

Limited Development

Limited development is an option that protects the majority of the land while a small portion is sold or maintained by the landowner for future development. In a limited

CULTURAL RESPECT EASEMENTS

A cultural respect easement is a legal agreement that guarantees Indigenous people cultural access to their native land in perpetuity, providing safe areas to practice traditional and spiritual lifeways, such as ceremonies, seasonal celebrations, camping, and more. These easements are placed on land that has been permanently protected from changes in land use, such as through a conservation easement or by transferring the land to a qualified conservation organization. Landowners interested in learning more about cultural respect easements should contact the local tribe or Indigenous-led organization.

development scenario, the areas with the greatest conservation value are protected through one of the tools previously discussed, while other less sensitive areas of the land are set aside for future development.

Finding a Conservation Organization

A conservation organization can help you explore these permanent land protection options. Many land conservation organizations seem exactly alike at first glance, but their missions and land management philosophies can vary greatly. Public conservation agencies and land trusts achieve their mission, in part, through land conservation. Your property's location, size, and natural resources all help determine which conservation organizations may be interested in working with you to conserve your land. If you do decide to move forward with land conservation, it is important that the organization you work with shares your goals and personal philosophy about land and land management. To find a land trust or public conservation agency working in your town, see Resources on page 264.

How Much Is This Going to Cost?

Estate planning is not cheap; however, your estate will also likely incur costs if you do nothing. Estate planning can help save money by reducing taxes. By planning in advance, you will have peace of mind in knowing that your land will be taken care of in the future. Costs go up the more indecisive you are and the more complex the solutions are, so try to come to as much of a decision as possible before bringing in more costly experts, like attorneys. Ask friends and neighbors for referrals, and don't be shy about asking professionals for an estimate before beginning to work with them.

TAXES

It has been said that the only sure things in life are death and taxes; estate planning is about preparing for both. Land is likely one of the most valuable assets in your estate, and due to the historic rise in real estate values, your land's value may be greater than you think.

The amount and types of taxes your estate may face depend on the value of your land, the form of ownership your land is held in, and how your assets—including your land—are transferred to your family. The aim, of course, is to pass on your assets in a way that meets your family's goals while minimizing the amount of taxes for which your estate becomes responsible.

Following is a summary of the various taxes that may affect you as a landowner as you move forward with estate planning. Remember that the laws that determine these taxes can change. The descriptions do not represent the law in any particular year; they are a simple explanation of the taxes that may be involved when land is transferred between people or generations of a family. Speak with a certified public accountant or a tax attorney who is familiar with land conservation for information specific to your situation. Understanding how permanent conservation can benefit you financially is a critical part of making an informed decision about the future ownership and use of your land.

Federal and state estate taxes. These are taxes your estate is subject to if its value exceeds a certain threshold. Federal and state tax thresholds often change from year to year. One way to lower the value of your estate is through land conservation tools.

Gift taxes. These are federal taxes incurred on gifts given while you are living. You can give a certain amount per year without triggering this tax. Giving under the taxable limit can be a useful way of transferring ownership or interest in your land slowly while avoiding taxes. If you give more than the limit annually, the excess is applied toward your lifetime gift-tax exclusion. If at any point the gifts you give during your lifetime or leave in your estate exceed that exclusion, the donor generally pays gift tax on the excess amount. Gifts of any amount transferred between spouses are allowed tax-free.

Capital gains taxes. These taxes are assessed when you sell capital assets, including land. Capital gains tax is applied to the appreciation value that your land and other assets have gained over time. For example, if you bought your land for $50,000 and it is now worth $200,000, the capital gains tax is applied to the increase in your land's value of $150,000. Placing a conservation restriction on your land is one effective way to lower its sale price and therefore the capital gain from the sale of your property. However, a bargain sale or the sale of a conservation easement can also trigger capital gains liability.

Federal income taxes. These taxes are based on income. Federal income tax can be reduced through a bargain sale, donation of land, or donation of a conservation easement by providing you with a charitable donation.

Property taxes. As land values and assessments increase, local property tax burdens can be difficult for families to meet. Conservation restrictions and current use tax programs provide opportunities to reduce property taxes on your land.

Working with a tax attorney or certified public accountant with experience in land conservation will help you understand the financial implications of your land conservation options. Maximizing the financial value of protecting your land often includes minimizing tax liability. A professional with experience can walk you through scenarios to inform your decision.

GETTING STARTED

The decisions (or lack of decisions) you make about your land will have financial and personal impacts long beyond your passing. Your land is an important but flexible asset that can be used to meet a wide variety of personal goals. Estate planning can be an intimidating and sometimes lengthy process, so don't delay—take a step today to begin the process. The hardest step is often the first. Identify questions you have, and reach out to a professional who can help you make an informed decision.

Work to formalize your wishes for your land. Don't leave the future of your land, and potentially family relationships, to chance. Once you have made decisions and created a plan, consider it a living document that will need to be revisited every 5 to 10 years or during major life changes.

FIELDWORK

Begin with Two Questions

The estate planning process need not feel overwhelming. The best way to approach this complicated task is one step at a time. There are two important questions that can help jump-start the planning process:

- Do you want to determine who will own your land?
- Do you want to control the use of your land in the future?

Using your responses to these questions, articulate your personal goals for the land and your financial needs. Engage people whose input you feel is important to the decision-making process to understand their goals and needs.

Once your goals are clearly articulated, contact a professional to help you achieve them. For example, if you are interested in keeping some or all of the forest as forest, reach out to a local land trust to find out your options. If you would like to investigate legal instruments, such as a will or a trust, contact an estate-planning attorney with experience in land conservation. Contact a forester to get enrolled in your state's current-use program.

CASE STUDY

Passing a Land Ethic to the Next Generation

The Riley Family

LOCATION: Barre, Massachusetts

ACRES: 280

Communicating your wishes for your land to your heirs is a critical first step in estate planning. Even more important, your wishes should be formally put into legal documents. In the case of Beatrice Riley, because she clearly and persistently communicated her wishes to her heirs, they worked hard to carry her wishes through, even in the absence of a legal imperative in her will.

Beatrice owned a house with 280 acres of land in Barre, Massachusetts. She loved the land and tried to instill a strong land ethic in her grandchildren by taking them blueberry picking, canoeing, fishing, and jeep riding on the trails. As the years passed, she would often talk to her family members about the land and ask them if they would care for it like she had.

When Beatrice grew older, two of her three grandchildren moved back to the house to take care of her. She became very ill, and when her death was imminent, the grandchildren needed to make a decision about what to do with the property. There was no direction in the will. Fortunately for Beatrice, one of her grandchildren who had power of attorney wanted to honor her grandmother's wishes to protect the land. Still, the family was concerned that the estate, capital gains, and other taxes would be too great to bear, forcing them to sell the land for development. All three grandchildren had a number of conversations about the future of the land and agreed that they wanted to honor their grandmother's wishes and conserve the land, but also hoped to be able to afford to live there with their families by somehow reducing the taxes associated with the inheritance.

The family attorney, who had had a longtime relationship with Beatrice, wasn't sure how best to help the grandchildren conserve the land, so he recommended that they contact Mount Grace Land Conservation Trust. The land trust helped the family understand their options and advised the family attorney. Because the Riley land had important natural resource values, the Massachusetts Department of Conservation and Recreation purchased 80 acres of it to protect the water quality of the Quabbin Reservoir.

To protect the remaining 200 acres, the family donated to the land trust a conservation easement (CE) that included all but an area around the original homestead and two other areas where the grandchildren could build houses in the future. Placing a CE on the property was also in their financial interest: the CE significantly reduced the value of the land and the taxes associated with the inheritance, and the donation of the CE was taken as a charitable gift and reduced the taxes even further. As a result, the three grandchildren and their families were able to afford to keep and live on the land. Even though the land surrounding the houses is permanently conserved, the CE allows them to continue to use the land for recreation, gardening, and forestry.

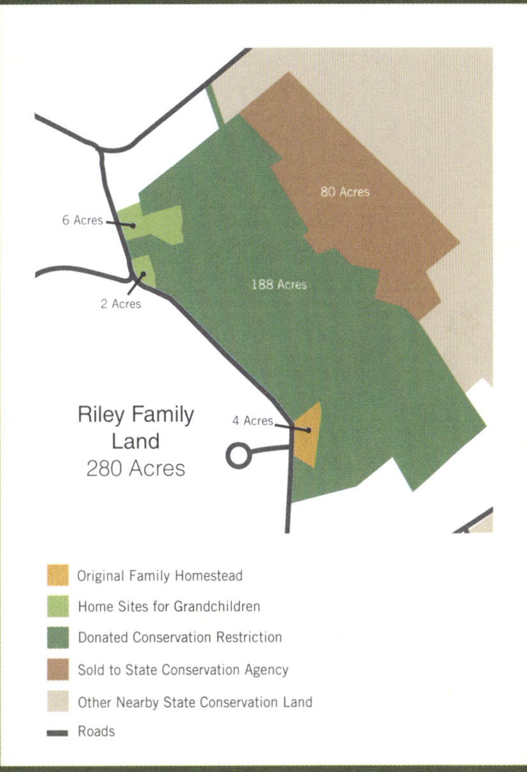

Communicating your wishes for your land to your heirs is a critical first step in estate planning.

Practicing Ecological Forestry for the Legacy of the Land

CONTINUITY: Without forests there are no forest benefits. Land is most at risk from parcellation and forest conversion at times of ownership transfer. Engaging in conservation-based estate planning will help ensure your forest stays forest through the ownership transition.

COMPLEXITY AND TIME: Forests need time—decades or centuries—to develop complexity and restore their missing cogs and wheels. Use conservation-based estate planning to ensure your forest has the time necessary for the development of structural complexity, including appropriate deadwood, decay in individual trees that leads to cavities, complex bark and crowns, and trees of large size.

CONTEXT: Low forest conversion rates and forest connectivity are critical characteristics of resilient landscapes. While conservation-based estate planning meets your personal and financial goals, it also adds to the resilience of your landscape by ensuring your forest will remain forested, providing an important anchor point within the landscape.

Epilogue

In his book *The Gift of Good Land*, author, philosopher, and farmer Wendell Berry describes the characteristics of a good solution. Berry writes that when health is named as the goal, then "a good solution causes a ramifying series of positive consequences." He reminds us that a good solution is in harmony with larger patterns.

We are in need of good solutions to protect our forests and the life-sustaining benefits they provide. Ecological forestry seeks to emulate the larger natural patterns of forests, providing solutions that result in a cascading series of positive ecological, social, and cultural consequences. We believe ecological forestry is a reconciliation of our relationship not only with forests but also with one another—a shift in culture that recognizes the value of forests, our dependence on them, and the role each of us plays in the relationship.

As a forest landowner, you stand in a critical and unique position to help. Practicing ecological forestry brings with it many benefits, including benefits to your forest, yourself, your family, and your community. Let your relationship and interaction with the forest grow. Trust that the connection you create will inform your decisions in unexpected ways. Once you've set out on the path forward, quite often the next step becomes obvious. Wonderful experiences await you.

It's time to start practicing ecological forestry.

ACKNOWLEDGMENTS

Science is a process. New ideas are laid on the foundation of those that came before. The threads of ecological forestry go back more than a century. Jerry Franklin and Norm Johnson pulled these threads together and gave them structure. Brian Palik continues to refine and promote ecological forestry and is gracious enough to have me (Tony D'Amato) along for the ride. This book uses their concepts and, hopefully, helps to increase the adoption of ecological forestry in our region.

The development of ideas and perspective is also a process. It takes knowledge, experience, mentorship, and time to bring an idea to maturity and to refine a perspective. Our ideas and perspective are the expression of all that we have learned from the many people we have encountered in our professional lives. Of special note are the following people: Mark Adams, Les Benedict, Karen Bennett, Mark Bodamer, Art Brooks, Brett Butler, Susan Campbell, Lina Clifford, Keystone Cooperators, Guthrie Diamond, Greg Edge, Kevin Evans, Arthur Eve, Tyler Everett, Wendy Ferris, Jennifer Fish, Shawn Fraver, Brad Hutnik, Maria Janowiak, Scott Jackson, Dave Kittredge, Bill Leak, Olivia Lukacic, Nancy Lyon, Amanda Mahaffey, Marla Markowsi-Lindsay, Massachusetts Woodlands Cooperative, Helena Murray, Linda Nagel, Native Land Conservancy, Ralph Nyland, Dave Orwig, Mike Parker, Klaus Puettmann, Jay Rasku, Norm Richards, John Scanlon, Bob Seymour, Emily Silver, Jim Soper, Sarah Wells, and Mariko Yamasaki.

There are several people who have been instrumental in the writing of this book. We thank Olivia Lukacic for contributing two case studies from her graduate work and an illustration, which were used in the book. We're appreciative of Les Benedict's contributions regarding Indigenous perspectives of forests. We are grateful to Michael Snyder for reviewing the book and providing thoughtful comments and suggestions that strengthened it. We would also like to thank Hannah Fries for her support, flexibility, and hard work in editing the book and bringing our vision to reality.

Finally, we would like to thank our families, Amy, Eli, and Adella Catanzaro and Jess Butler and Quinn D'Amato, for their support as we pursue our passion of forest conservation, and for their patience as we talk about deadwood.

Resources

Contact a Professional in Your State

CONNECTICUT

Connecticut Department of Energy and Environmental Protection (DEEP), Forestry
https://portal.ct.gov/deep/forestry
Find your service forester
Find a consulting forester
Information about forests and forestry programs

UConn Extension Forestry
https://ctforestry.cahnr.uconn.edu

Connecticut Land Conservation Council
https://ctconservation.org/find-a-land-trust
Find a land trust

DELAWARE

Delaware Department of Agriculture, Forest Service
https://agriculture.delaware.gov/forest-service
Find your service forester
Find a consulting forester
Information about forests and forestry programs

University of Delaware, Cooperative Extension, Forestry
https://udel.edu/academics/colleges/canr/cooperative-extension/environmental-stewardship/forestry

Land Trust Alliance
https://landtrustalliance.org/land-trusts/gaining-ground/delaware
Find a land trust

ILLINOIS

Illinois Department of Natural Resources, Forestry Resources
https://dnr.illinois.gov/conservation/forestry
Find your service forester
Find a consulting forester
Information about forests and forestry programs

University of Illinois Extension, Forestry
https://extension.illinois.edu/forestry/programs

Prairie State Conservation Coalition
https://prairiestateconservation.org
Find a land trust

Illinois Call Before You Cut
https://callb4ucut.com/illinois

INDIANA

Indiana Department of Natural Resources, Division of Forestry
https://in.gov/dnr/forestry
Find your service forester
Information about forests and forestry programs

Find an Indiana Forester
https://findindianaforester.org

Purdue University Extension
https://extension.purdue.edu

Women4theLand
https://women4theland.org

Indiana Land Protection Alliance
https://protectindianaland.org
Find a land trust

Indiana Call Before You Cut
https://callb4ucut.com/indiana

IOWA

Iowa Department of Natural Resources, Forestry Resources
https://iowadnr.gov/programs-services/forestry-resources
Find your service forester
Information about forests and forestry programs

Iowa State University Extension and Outreach, Natural Resource Stewardship
https://naturalresources.extension.iastate.edu

Iowa Landowner Options
https://iowalandoptions.org
Find a land trust

Iowa Call Before You Cut
https://callb4ucut.com/iowa

MAINE

Maine Department of Agriculture, Conservation & Forestry, Maine Forest Service
https://maine.gov/dacf/mfs
Find your service forester
Find a consulting forester
Information about forests and forestry

University of Maine Cooperative Extension, Taking Care of Your Woodland
https://extension.umaine.edu/woodland

Maine Land Trust Network
https://mltn.org/trusts
Find a land trust

MARYLAND

Maryland Department of Natural Resources, Forest Service
https://dnr.maryland.gov/forests
Find your service forester
Information about forests and forestry
Find a land trust

University of Maryland Extension, Woodland Stewardship Education
https://extension.umd.edu/woodland
Find a consulting forester
Information about forests and forestry

Maryland Call Before You Cut
https://callb4ucut.com/maryland

MASSACHUSETTS

Massachusetts Bureau of Forest Fire Control and Forestry
https://mass.gov/orgs/bureau-of-forest-fire-control-and-forestry
Find your service forester
Information about forests and forestry

UMassAmherst (MassWoods)
https://masswoods.org
Find your service forester
Find a consulting forester
Find a land trust
Information about forests and forestry

Massachusetts Land Trust Coalition
https://massland.org
Find a land trust

MICHIGAN

Michigan Department of Natural Resources, Forestry
https://michigan.gov/forestry
Find your service forester
Find a consulting forester
Information about forests and forestry programs

Michigan Department of Agriculture and Rural Development, Private Forestlands Initiative
https://michigan.gov/mdard/environment/forestry

Michigan State University Extension, Forestry
https://canr.msu.edu/forestry

Land Trust Alliance
https://landtrustalliance.org/land-trusts
Find a land trust

Michigan Call Before You Cut
https://callb4ucut.com/michigan

MINNESOTA

Minnesota Department of Natural Resources, Forestry Division
https://dnr.state.mn.us/forestry
Find your service forester
Find a consulting forester

University of Minnesota Extension, My Minnesota Woods
https://extension.umn.edu/natural-resources/my-minnesota-woods

Land Trust Alliance
https://landtrustalliance.org/land-trusts
Find a land trust

Minnesota Women's Woodland Network
https://mnwwn.org

Minnesota Call Before You Cut
https://callb4ucut.com/minnesota

MISSOURI

Missouri Department of Conservation, Trees & Plants
https://mdc.mo.gov/trees-plants
Find your service forester
Information on forests and forestry programs

Missouri Consulting Foresters Association
https://missouriforesters.com
Find a consulting forester

University of Missouri Extension, Forest Ecology and Management
https://extension.missouri.edu/programs/forest-ecology-and-management

Land Trust Alliance
https://landtrustalliance.org/land-trusts
Find a land trust

Missouri Call Before You Cut
https://callb4ucut.com/missouri

NEW HAMPSHIRE

New Hampshire Department of Natural and Cultural Resources, New Hampshire Forests and Lands
https://nhdfl.dncr.nh.gov

University of New Hampshire Extension, Natural Resources
https://extension.unh.edu/natural-resources
Find your service forester
Find a consulting forester

New Hampshire Land Trust Coalition
https://nhltc.org
Find a land trust

NEW JERSEY

New Jersey Department of Environmental Protection, New Jersey Forest Service
https://nj.gov/dep/parksandforests/forest
Find your service forester
Information on forests and forestry programs

Rutgers Cooperative Extension, New Jersey Agricultural Experiment Station, Department of Agriculture and Natural Resources
https://njaes.rutgers.edu/anr

Land Trust Alliance
https://landtrustalliance.org/land-trusts
Find a land trust

NEW YORK

New York Department of Environmental Conservation, Division of Lands and Forests
https://dec.ny.gov/nature/forests-trees
Find your service forester
Find a consulting forester
Information on forests and forestry programs

Cornell University Cooperative Extension, ForestConnect
https://blogs.cornell.edu/cceforestconnect

Land Trust Alliance
https://landtrustalliance.org/land-trusts
Find a land trust

OHIO

Ohio Department of Natural Resources (ODNR), Division of Forestry
https://ohiodnr.gov/discover-and-learn/safety-conservation/about-ODNR/forestry
Find your service forester
Find a consulting forester
Information on forests and forestry programs

Ohio State University Extension, School of Environment and Natural Resources (SENR)
https://senr.osu.edu

Coalition of Ohio Land Trusts
https://ohiolandtrusts.org
Find a land trust

Ohio Call Before You Cut
https://callb4ucut.com/ohio

PENNSYLVANIA

Pennsylvania Department of Conservation and Natural Resources, Bureau of Forestry
https://pa.gov/agencies/dcnr/programs-and-services/about/bureaus-and-offices/forestry
Find your service forester
Find a consulting forester
Information on forests and forestry programs

Penn State Extension
https://extension.psu.edu

Land Trust Alliance
https://landtrustalliance.org/land-trusts
Find a land trust

RHODE ISLAND

Rhode Island Department of Environmental Management (DEM), Forest Environment
https://dem.ri.gov/programs/forestry
Find your service forester
Find a consulting forester
Information on forests and forestry programs

University of Rhode Island Cooperative Extension (Rhode Island Woods)
https://web.uri.edu/rhodeislandwoods

Rhode Island Land Trust Council
https:// rilandtrusts.org
Find a land trust

VERMONT

Vermont Department of Forests, Parks, and Recreation
https://fpr.vermont.gov/forests
Find your service forester
Find a consulting forester
Information on forests and forestry

University of Vermont Extension (Our Vermont Woods)
https://ourvermontwoods.org

Land Trust Alliance
https://landtrustalliance.org/land-trusts
Find a land trust

WEST VIRGINIA

West Virginia Division of Forestry
https://wvforestry.com
Find your service forester
Find a consulting forester
Information on forests and forestry

West Virginia University Extension
https://extension.wvu.edu

Land Trust Alliance
https://landtrustalliance.org/land-trusts
Find a land trust

West Virginia Call Before You Cut
https://callb4ucut.com/westvirginia

WISCONSIN

Wisconsin Department of Natural Resources, Forestry Division
https://dnr.wisconsin.gov/topic/forestry
Find your service forester
Find a consulting forester
Information on forests and forestry programs

University of Wisconsin–Madison Division of Extension, Forestry
https://woodlandinfo.org

Wisconsin Woodland Owners Association, Women of WWOA
https://wisconsinwoodlands.org/women-of-wwoa

Gathering Waters: Wisconsin's Alliance for Land Trusts
https://gatheringwaters.org
Find a land trust

General Resources

FOREST RESILIENCE

Northern Research Station Climate Change Atlas, Tree Atlas
https://fs.usda.gov/nrs/atlas/tree
Find detailed projections for species in your region.

Climate Change Response Framework
https://forestadaptation.org

FOREST LANDOWNER ORGANIZATIONS

Most states have a statewide landowner organization. These organizations provide excellent pathways to meet other landowners and learn about educational opportunities. To learn more about your specific state's forest landowner organization, contact your local service forester.

Women Owning Woodlands
https://womenowningwoodlands.org

American Forest Foundation
https://forestfoundation.org

American Tree Farm System
https://treefarmsystem.org

FEDERAL PROGRAMS

Forest Legacy Program (FLP)
This USDA Forest Service program encourages privately owned forests through perpetual conservation easements or land purchases, encouraging sustainable forest management through forest land protection. Contact your state forestry agency for more information. See the Contact a Professional in Your State section on page 264.

Forest Stewardship Program (FSP)
This USDA Forest Service program offers resources to help landowners interested in active stewardship develop a management plan. Contact your state forestry agency for more information. See the Contact a Professional in Your State section on page 264.

Environmental Quality Incentives Program (EQIP)
This USDA Natural Resources Conservation Service (NRCS) program provides financial and technical assistance to address concerns regarding environmental and natural resources. The program is funded through the Farm Bill. Contact your local NRCS office to learn more: *https://nrcs.usda.gov/contact/find-a-service-center*

PROFESSIONAL ORGANIZATIONS FOR FORESTERS

Association of Consulting Foresters
https://acf-foresters.org

Forest Stewards Guild
https://foreststewardsguild.org

Society of American Foresters
https://eforester.org

Source Material by Chapter

Introduction

Kimmerer, Robin Wall. 2015. *Braiding Sweetgrass*. Milkweed Editions.

Leopold, Aldo. 2020. *A Sand County Almanac*. Oxford University Press.

CHAPTER 1: How Your Forest Works

Barnes, B. V., D. R. Zak, S. R. Denton, and S. H. Spurr. 1998. *Forest Ecology*, 4th edition. Wiley.

CHAPTER 2: Forests of the Past

Foster, David R. , and John F. O'Keefe. 2000. *New England Forests Through Time: Insights from the Harvard Forest Dioramas*. Harvard University Press.

Schulte, L. A., D. J. Mladenoff, T. R. Crow, L. C. Merrick, and D. T. Cleland. 2007. "Homogenization of Northern US Great Lakes Forests Due to Land Use." *Landscape Ecology* 22: 1089–1103.

CHAPTER 3: Forests of Today

New England and Northern New York Forest Ecosystem Vulnerability Assessment and Synthesis: A Report from the New England Climate Change Response Framework Project. 2018. USDA Forest Service Northern Research Station. General Technical Report NRS-173.

CHAPTER 4: Tending the Future with Ecological Forestry

D'Amato, Anthony W., Brian J. Palik, Jerry F. Franklin, and David R. Foster. 2017. "Exploring the Origins of Ecological Forestry in North America." *Journal of Forestry* 115 (2): 126–127. https://doi.org/10.5849/jof.16-013.

Franklin, Jerry F., K. Norman Johnson, and Debora L. Johnson. 2018. *Ecological Forest Management*. Waveland Press.

Palik, Brian J., Anthony W. D'Amato, Jerry F. Franklin, and K. Norman Johnson. 2021. *Ecological Silviculture*. Waveland Press.

CHAPTER 5: Choosing Your Stewardship Approach

Lukacic, O., P. Catanzaro, E. Huff, and K. Hamunen. 2021. *Women on the Land.* University of Massachusetts, Cooperative Extension Landowner Outreach Publication, 38 pp.

CHAPTER 6: Restoring Old-Growth Characteristics

D'Amato, A., and P. Catanzaro. 2023. *Restoring Old-growth Characteristics to New England's and New York's Forests.* University of Massachusetts, Cooperative Extension Landowner Outreach Publication, 34 pp.

CHAPTER 7: Mitigating Climate Change

Catanzaro, P., and A. D'Amato. 2019. *Forest Carbon: An Essential Natural Solution for Climate Change.* University of Massachusetts, Cooperative Extension Landowner Outreach Pamphlet, 28 pp.

Hoover, Coeli M., Ben Bagdon, and Aaron Gagnon. 2021. "Standard Estimates of Forest Ecosystem Carbon for Forest Types of the United States." General Technical Report NRS-202. Madison, WI: US Department of Agriculture, Forest Service, Northern Research Station. 158 pp. https://doi.org/10.2737/NRS-GTR-202.

CHAPTER 8: Diversifying Wildlife Habitat

DeGraaf, Richard M., and Deborah D. Rudis. 1986. "New England Wildlife: Habitat, Natural History, and Distribution." General Technical Report NE-108. US Department of Agriculture, Forest Service, Northeastern Forest Experimental Station, 491 pp. https://doi.org/10.2737/NE-GTR-108.

DeGraaf, Richard M., Mariko Yamasaki, William B. Leak, and Anna M. Lester. 2005. *Landowner's Guide to Wildlife Habitat: Forest Management for the New England Region.* University of Vermont Press and University Press of New England. https://www.fs.usda.gov/nrs/pubs/jrnl/2005/ne_2005_degraaf_001.pdf.

CHAPTER 9: Building Forest Resilience

Catanzaro, P., A. D'Amato, D. Orwig, N. Siegert, L. Benedict, T. Everett, J. Daigle, and A. Mahaffey. *Managing Northeastern Forests Threatened by Emerald Ash Borer*. University of Massachusetts, Cooperative Extension Landowner Outreach Publication, 30 pp.

Catanzaro, P. F., A. D'Amato, and E. Silver Huff. 2016. *Increasing Forest Resiliency for an Uncertain Future.* University of Massachusetts, Cooperative Extension Landowner Outreach Pamphlet, 30 pp.

Climate Change Response Framework: https://forestadaptation.org

Cornell University Slash Wall Resource Center: https://blogs.cornell.edu/slashwall

Rawinski, Thomas J. 2014. *White-tailed Deer in Northeastern Forests: Understanding and Assessing Impacts.* Report NA-IN-02-14. US Department of Agriculture, Forest Service, Northeastern Area State and Private Forestry.

USDA Forest Service Climate Change Atlas, Tree Atlas: https://fs.usda.gov/nrs/atlas/tree

CHAPTER 10: Local Wood, Local Good

Berlik, M., D. Kittredge, and D. Foster. 2002. *The Illusion of Preservation.* Harvard Forest Paper No. 26.

Guide for Planning the Local Victory Garden Program. 1942. United States Office of Civilian Defense, USDA.

Littlefield, C., B. Donahue, P. Catanzaro, D. Foster, A. D'Amato, K. Laustsen, and B. Hall. 2024. *Beyond the "Illusion of Preservation."* University of Massachusetts, Cooperative Extension Landowner Outreach Publication, 48 pp.

Ryan, T. Personal communication. March 2024.

CHAPTER 11: The Successful Timber Harvest

Catanzaro, P. F., and D. B. Kittredge. 2004. *Common Elements of a Timber Sale Contract.* University of Massachusetts, Cooperative Extension Outreach Document, 4 pp.

CHAPTER 12: The Legacy of Your Land

Catanzaro, P. F., and W. Sweetser Ferris, J. Rasku, and S. Wells. 2017. *Protecting Your Legacy*. University of Massachusetts, Cooperative Extension Landowner Outreach Pamphlet, 20 pp.

Index

Page numbers in *italics* indicate illustrations or photos; numbers in **bold** indicate charts.

D

E

P

R

S